SpringerBriefs in Applied Sciences and Technology

SpringerBriefs present concise summaries of cutting-edge research and practical applications across a wide spectrum of fields. Featuring compact volumes of 50 to 125 pages, the series covers a range of content from professional to academic.

Typical publications can be:

- A timely report of state-of-the art methods
- An introduction to or a manual for the application of mathematical or computer techniques
- A bridge between new research results, as published in journal articles
- A snapshot of a hot or emerging topic
- An in-depth case study
- A presentation of core concepts that students must understand in order to make independent contributions

SpringerBriefs are characterized by fast, global electronic dissemination, standard publishing contracts, standardized manuscript preparation and formatting guidelines, and expedited production schedules.

On the one hand, **SpringerBriefs in Applied Sciences and Technology** are devoted to the publication of fundamentals and applications within the different classical engineering disciplines as well as in interdisciplinary fields that recently emerged between these areas. On the other hand, as the boundary separating fundamental research and applied technology is more and more dissolving, this series is particularly open to trans-disciplinary topics between fundamental science and engineering.

Indexed by EI-Compendex, SCOPUS and Springerlink.

Silvio Vaz Jr.

The Lignin Macromolecule

A Compendium of Sustainable Technologies

 Springer

Silvio Vaz Jr.
American Chemical Society, and
Graduate Program of Biofuels-Federal
University of Uberlândia
Brasília, Brazil

ISSN 2191-530X   ISSN 2191-5318 (electronic)
SpringerBriefs in Applied Sciences and Technology
ISBN 978-3-031-75510-1   ISBN 978-3-031-75511-8 (eBook)
https://doi.org/10.1007/978-3-031-75511-8

This Springer imprint is published by the registered company Springer Nature Switzerland AG
The registered company address is: Gewerbestrasse 11, 6330 Cham, Switzerland

If disposing of this product, please recycle the paper.

I would like to dedicate this book to my father Silvio Vaz, my mother Isaura Carvalho Vaz, and to my aunt Natalia Vaz. In memorian, and always with me.

Preface

Lignun is the Latin word for wood. It represents the strong and the relevance of lignocellulosic biomass as the primordial source of materials and energy for the humankind.

When we obtain lignin from the lignocellulosic biomass, we have a prodigious carbon-based renewable raw material for several products as chemicals, fuels, and end-use materials with the possibility to generate wealth under the sustainability concept. Furthermore, the lignin usages promote technological and scientific advances anchored by environmental gains.

From these statements, this book proposes to explore the sustainable usages and applications of lignin, based on a strong holistic view of production and value chains.

The author carried out a careful survey of the main industrial technologies in use, according to his academic and professional experiences. Moreover, author also considered trends observed in current R&D&I scenarios, regulatory legislation, market, and other relevant aspects, highlighting the principles of green chemistry.

Eight chapters compose the book: Chapter 1 deals with an overview of the lignin production, its global availability, market, and products and applications which are treated in order to introduce its relevance as an industrial renewable macromolecule with several uses and opportunities to be explored.

Chapter 2 describes the physicochemical considerations of the macromolecule, metabolic pathways related to the lignin biosynthesis, and differences between hardwoods and softwoods in order to understand the biochemical aspects behind the natural production of lignin.

Chapter 3 treats about the description of the most used advanced analytical chemistry, a critical application of the analytical techniques for Kraft lignin as a case study, and new analytical approaches in order to generate analytical information for the macromolecule understanding and usages.

Chapter 4 discusses the main extraction process to obtain lignin-types native, Kraft, Organosolv, soda and Klason, and recent advances as LignoBoost and Ligno-Force. Additionally, Kraft and Organosolv processes are evaluated according to green chemistry principles. And the LignoBoost process is presented as a case study.

Chapter 5 deals with the classes of chemical conversion processes in order to explore all possibilities to achieve several organic compounds. A case study base on lignin cracking is presented. And the evaluation of advantages and disadvantages according the green chemistry principles is considered.

Chapter 6 treats about enzymatic catalysis and microbes (fungi, bacteria, yeasts) application to produce several organic compounds. The vanillin production by fungi and the green chemistry principles application to bioprocesses are also explored.

Chapter 7 deals with new products as battery components, polymeric materials, additive for pellets, and carbon fiber in order to demonstrate the technological opportunities for these lignin-based materials. Furthermore, a case study will provide the potential of carbon fibers, and the evaluation of advantages and disadvantages according to green chemistry principles.

Finally, Chap. 8 deals with the concepts of circularity and sustainability allied to bio-economy, life cycle assessment, and industrial ecology in order to promote sustainable technologies based on lignin. Additionally, know the biobased content is paramount to guarantee the origin and the sustainability of products and processes, with will be possible by means radiocarbon analysis. Furthermore, the E-factor and the decarbonization promoted by the lignin-based carbon fibers are discussed as a case study.

Thus, the book offers to the reader information based on the start of the art and practical applicability of its content in order to stimulate as a pivotal component to achieve a more sustainable and green applied chemistry.

Good lecture!

Brasília, Brazil   Silvio Vaz Jr.
2024

Contents

Chapter 1
Introduction to the Lignin Subject

Abstract Lignin, a macromolecule obtained from lignocellulosic biomass, is a residue or coproduct from industrial sectors as cellulose pulp and paper and second-generation ethanol. It is classified as sulfur containing lignin and sulfur-free lignin, that depends on the extraction process. In this chapter aspects related to an overview of the lignin production, its global availability, market, and products and applications are treated in order to introduce the relevance of lignin as an industrial renewable macromolecule with several usages.

Keywords Renewable macromolecule · Sulfur contain lignin · Sulfur-free lignin · Global availability · Bioproducts

1.1 Lignin as a Renewable Macromolecule

Lignin is as phenolic-based macromolecule—sometimes known as a biopolymer—obtained from plant biomass (Fig. 1.1) and which its name is originated from the Latin word *lignum* that means wood. Generally, it is a residue or a coproduct from biobased industrial sectors as cellulose pulp and paper (the main source) and second-generation ethanol (or cellulosic ethanol). That means we need a previous step of extraction to release lignin from another lignocellulosic biomass fractions or components—i.e., cellulose and hemicellulose. It is worth to note that wood can be visualized as a composite of cellulose fibers embedded in lignin.

The lignin global production and availability—discussed in this chapter—depends on the presence of these types of industries and agroindustrial biomass (e.g., eucalyptus, pinus, sugarcane) in a certain country or region, what will facilitate or will reduce the technological uses of the macromolecule.

Fig. 1.1 A typical representation of the lignin chemical structure (**a**)—we can see a high number of aromatic structures associated with oxygen-derived functional groups, as hydroxyl and ethers, e.g., phenol and methoxyl. And lignin powder obtained by the Kraft processing (**b**)

1.2   An Overview of the Lignin Production

Generally, the lignin production is classified from the presence or absence of sulfur, based on the pulping process [1]:

Sulfur containing lignin:

- Kraft lignin (S content: 1.0—3.0 wt%), obtained from pulp and paper production.
- Lignosulfonates (S content: 3.5—8.0 wt%), obtained from chemicals production.
- Hydrolyzed lignin (S content: 0.0—1.0 wt%), obtained from the second-generation (or cellulosic) ethanol production.

Sulfur-free lignin:

- Organosolv lignin, from the lignocellulosic biomass fractionation.
- Soda lignin, obtained from pulp production.
- Lignin from second-generation ethanol production.

The sulfur content comes from the extraction process to separate lignin from cellulose and hemicellulose, another lignocellulosic constituent from plant cell wall. One example is the Kraft lignin, which makes use of sodium sulfide associated with sodium hydroxide for the wood cooking in order to release the lignin macromolecule during the cellulose production for pulp and paper—the processes to produce each one is discussed in Chap. 4.

Lignin can be understood as a residue or as a coproduct from the lignocellulosic biomass processing according to its use. In this context, a residue is a substance without a direct or well-defined use (e.g., lignin present in the black liquor), whereas the coproduct is a substance with a secondary use (e.g., energy source) when compared to the product (e.g., cellulose and paper). Figures 1.2 and 1.3 present the process to obtain lignin from wood and sugarcane in order to obtain industrial raw material.

1.3 Lignin Global Availability

The global availability of lignin, thinking it as a potential chemical commodity for several uses and applications, is shown in Fig. 1.4. Despite a global distribution—except for Oceania—it is limited to the lignosulfonate, as a technical lignin (a lignin with a well-attributed purity grade and available for distribution and marketing).

Countries as USA and China, followed by Russia, South Africa, Canada, and Scandinavian countries have the main lignin supply capacity; Spain, Germany, France, Brazil, and India have a lower supply capacity.

From the lignin sources, we should consider the paper and pulp and cellulosic ethanol—major from sugarcane—sectors as the main actors due to well-established production and distribution chains. For both sectors, the global productions are 171 million metric tons for pulp and paper from virgin wood species [3] and 30 billion gallons per year (here considering first and second-generation ethanol) [4].

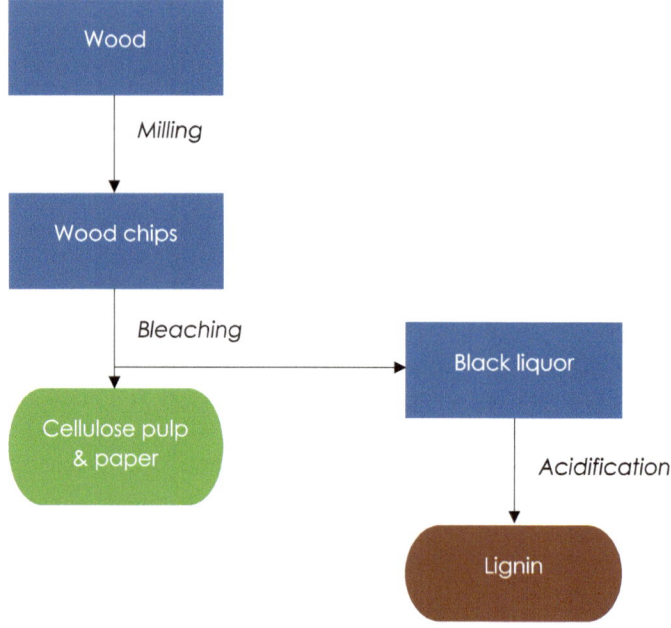

Fig. 1.2 Simplified flowchart for the obtention of lignin from wood by means the Kraft processing

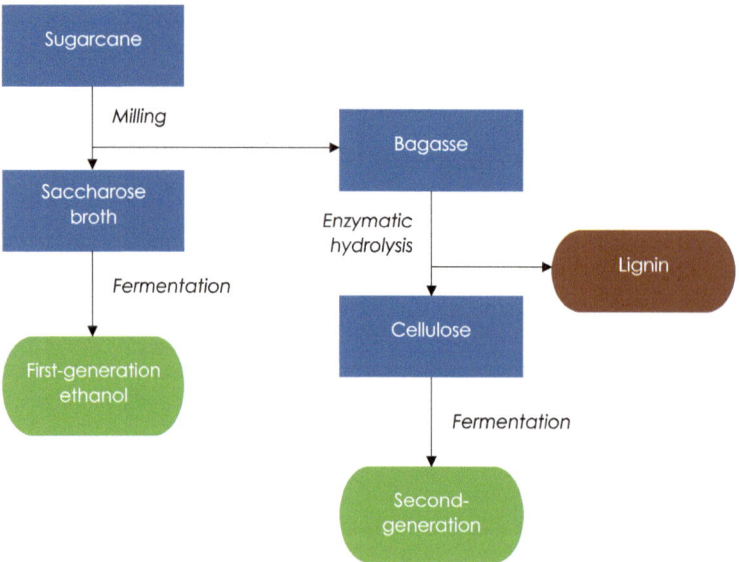

Fig. 1.3 Simplified flowchart for the obtention of lignin from sugarcane bagasse by means the enzymatic hydrolysis

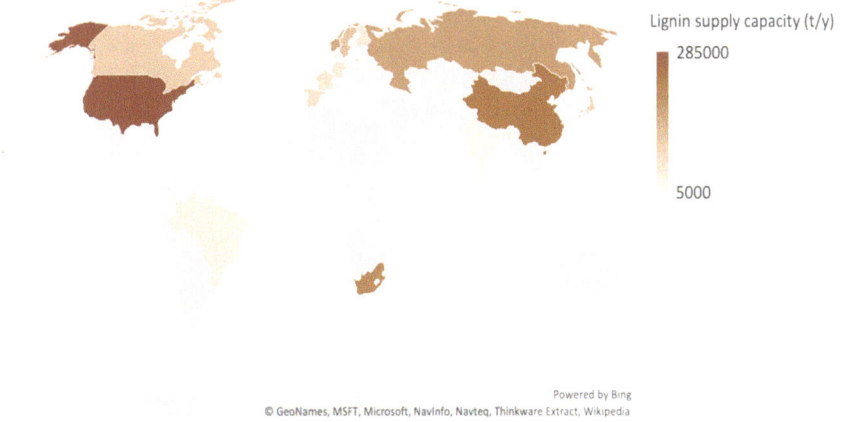

Fig. 1.4 Technical lignins' production capacity per country (as lignosulfonate), base years 2017–2018. *Source* Dessbesell et al. [2]. Reprinted with permission from Elsevier

1.4 Market Aspects

The global lignin market size was estimated at USD 1.08 billion in 2023 and is expected to grow at a compound annual growth rate (CAGR) of 4.5% from 2024 to 2030 [5]. Increasing demand for lignin in animal feed and natural products is anticipated to drive growth. Nevertheless, the lignin industry faced a backlash owing to disruptions in the value chain, including workforce losses, raw material supply, trade and logistics, and uncertain consumer demand.

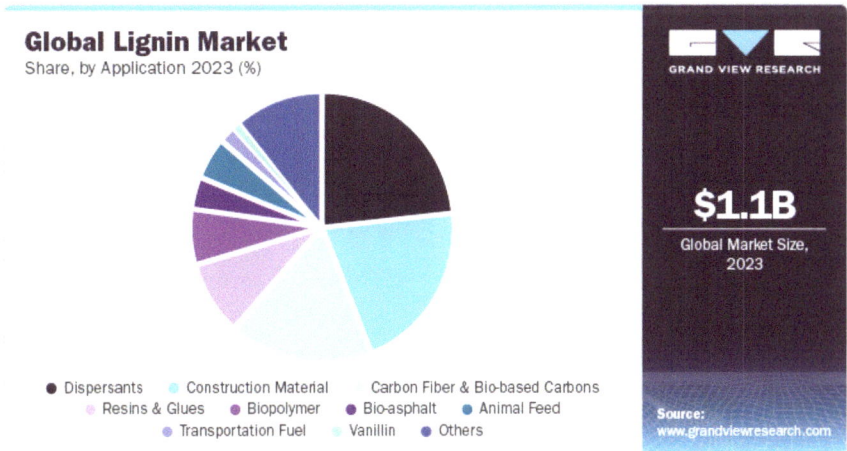

Fig. 1.5 Distribution of the global lignin market. Reprinted with permission from grand view research

Figure 1.5 depicts the composition of the global lignin market.

Dispersants (e.g., lignosulfonates), construction material, and carbon fiber and biobased carbons are the main market components, followed by animal feed, resins and glues, biopolymer, and bioasphalt. Transportation fuel, vanillin, and others are fewer representative components.

1.5   An Outlook of (Bio)products and Applications

According the International Lignin Institute [6], a task force addressed to spread the technological applications of lignin, and lignin-related products—or *bioproducts*—are:

- Multi-polarity-related products based on hydrophilic and hydrophobic groups, to be applied in dispersants (e.g., ceramics, oil well drilling, clay bricks and tiles, electrolytes), emulsions (e.g., wax, asphalt, vitamins, micronutrients), and others (e.g., complexing agents, flocculating, ion exchange, protein coagulants, destabilization of oil emulsions, corrosion protection).
- Materials based on natural branched and crosslinked network, to be applied as resins (e.g., phenolic resins, polyurethanes, epoxies), rubber reinforcing, composites, polyolefin, carbon sieves, activated carbon, carbon fibers, heat resistance, antioxidants, anti-inflammation.
- Agrochemicals for soil functionalization, plant and animal nutrition, to be applied as soil rehabilitation, slow release fertilizers, artificial humus, fertilizer encapsulation, humus improvement, soil stabilization, insecticides, granulation, pelletizing, and chelates.
- High purity/value applications, mainly substances with antioxidant, antibacterial and antiviral properties (e.g., antibacterial agents, HIV inhibitor, antioxidants, oxygen scavengers), besides for digestion regulation, plant immunology/growth stimulators, and hydrogels.
- And as miscellaneous, the application to energy production, binders, absorbents, etc.

These products and applications will be explored in depth in Chaps. 5, 6, and 7. Some companies offering lignin and lignin derivatives as:

- Borregaard (Norway)[1]: lignin and lignosulfonates to be used as binding agents, dispersing agents, crystal growth modifiers, emulsion stabilizers, and complexing agents.
- Suzano (Brazil)[2]: lignin for polymeric materials, cosmetics, feed, carbon materials, agricultural dispersing agents and products for civil construction.

---

[1] https://www.borregaard.com/product-areas/lignin/.

[2] https://www.suzano.com.br/en/products-and-brands/raw-material/lignin.

- Ingevity (USA)[3]: Kraft lignin to be used as emulsifiers, batteries components, soil stabilization, and remediation agents.

Regarding extraction technologies, Valmet (Finland) offers the LignoBoost process[4] for lignin extraction from pulp mill black liquor from pulp and paper industries.

However, the development of new products is an activity where the limit is almost the sky, as depicted in Table 1.1 just for chemical conversions.

Products presented in Table 1.1 can be considered, in some cases, as potential substitutes for petrochemicals, boosting the possibility to change a non-renewable fossil raw material to a renewable biobased raw material. And this approach is closely related to the principle 7 of green chemistry (renewable feedstock) [8] and to the sustainable development goal 11 for the responsible consumption and production [9].

Thus, we can observe a large application opportunity for lignin-based products, which will reflect, for instance, in a more sustainable chemical industry as the main enabler of the lignin-based technologies, as reported by Vaz Jr. and de Souza [10] for the opportunities of sustainable carbonaceous raw materials to reduce the negative impacts of chemical industry.

1.6 Conclusions

Lignin, a macromolecule obtained from lignocellulosic biomass, is a residue or coproduct from industrial sectors as pulp and paper and second-generation ethanol. It is classified as sulfur containing lignin and sulfur-free lignin, that depends on the extraction process.

Lignin has a global distribution, which the main lignin supply capacity—related to lignosulfonate—is associated with USA and China. However, European, Asian, and South American countries are also important players. And regarding the market aspects, it is expected a GAGR of 4.5% until 2030.

Finally, an outlook of lignin products and their applications can show a diversity of technological opportunities comprising, for instance, dispersants, emulsions, resins, composites, agrochemicals, and specialties.

---

[3] https://www.ingevity.com/featured-products/indulin/.

[4] https://www.valmet.com/pulp/other-value-adding-processes/lignin-extraction/lignoboost-pro cess/.

Table 1.1 High-value chemicals obtained from lignin depolymerization

Type of reaction	Types of lignin, catalysts, and reaction conditions	Products
Hydrogenolysis	Kraft lignin, Ni-Re/ Nb_2O_5 catalyst (mass ration 2.5) in ethanol as cosolvent 330 °C, 3 h	96.70 and 35.41 wt% of oil and lignin monomer yields
Directional depolymerization	Binary solvent system containing methanol-dimethoxymethane, $H_4SiW_{12}O_{40}$ catalyst, 200 °C, 1 h	Liquefied lignin (67.39 wt%) and monophenolics (27.67 wt%)
Microwave-assisted depolymerization	Dealkaline lignin, silicon carbide, dichloromethane vapor, 1000 W; 700 °C, 15 min	Aryl monomers and carbon nanospheres
Oxidative catalytic fractionation and depolymerization	Lignin fractionation, α-OH methyoxylation of lignin using polyoxometalate catalyst at 100 °C in methanol and water (9:1 v/v), and depolymerization with polyoxometalate catalyst at 140 °C	Depolymerized lignin (74 wt%) containing aromatic monomers (45.9 wt%)
Oxidative depolymerization	Organosolv lignin, nano WO_3, tert-butyl hydrogen peroxide, NaOH, 1, 4-dioxane, ethylene glycol, 140 °C, 4 h	Lignin oil (80.4 wt%) containing vanillic acid as a major aromatic monomer
Photocatalytic depolymerization	Dioxanesolv lignin, $Zn_4In_2S_7$/ graphene oxide-based photocatalyst in acetone	Phenols and ketones
Catalytic depolymerization	Alkali lignin, activated carbon-supported nickel-tungsten carbide (4%Ni-30%W_2C/AC) as catalyst, composite alkali (NaOH + Na_2SO_3 (3:1)), 260 °C, 5 h	Depolymerized lignin (94.4 wt%) containing aromatic monomers (17.18 wt%)
Catalytic hydrogenolysis	Demethylated lignin, Ru/C catalyst, isopropanol, 300 psi with He, 300 °C, 3 h	Catechol- and catechol derivatives-rich liquid
Catalytic depolymerization	Formic acid: lignin (mass ratio 4:1), Fe − Zn/Al_2O_3 catalyst in water, 180 °C, 6 h	Lignin oil (28.31 wt%) containing aromatic monomers
Oxidative depolymerization	Rice straw alkali lignin, Acidic catalyst ZSM-5, H_2O_2 under N_2, 160 °C, 1 h	Lignin oil containing 2-methoxy-4-vinylphenol as a major component along with vanillin and acetosyringone, etc.
Catalytic depolymerization	Kraft lignin, phosphorus modified molybdenum/sepiolite catalyst in ethanol, 7.5 MPa N_2, 290 °C, 4 h, 480 rpm	Lignin oil containing guaiacol as major component
Microwave and alkali-assisted depolymerization	Wheat straw lignin, NaOH in water 170 °C, 20 min	Depolymerized lignin (21.68 wt%) suitable for adhesive preparation

(continued)

Table 1.1 (continued)

Type of reaction	Types of lignin, catalysts, and reaction conditions	Products
Catalytic liquefaction	Kraft lignin, Re/Al$_2$O$_3$ catalyst, NaOH, hexadecane, H$_2$ 30 bar, 400 °C, 6 h	Monocyclic hydrocarbons and Monocyclic aromatic hydrocarbons
Catalytic hydrogenolysis	3% Cu@MIL-101(Cr) catalyst, 250 °C, 2.5 MPa H$_2$,6 h followed by 20 min of sonication	Depolymerized lignin with aromatic monomers (38.5 wt%)
Photocatalytic depolymerization	Organosolv lignin in dioxane, Bi-doped ZnO nanocomposites, 1 h under direct sun light	Depolymerized lignin (97 wt%) containing phenolic monomers
Photocatalytic depolymerization	Alkali lignin, carbon quantum dots decorated TiO$_2$ nanocomposite under direct sunlight, 6 h	99 wt% depolymerization of lignin containing *m*-anisic acid (3-methoxybenzoic acid) and *p*-hydroxybenzoic acid
Microwave-assisted hydrogenation/ hydrodeoxygenation	Dealkaline lignin, HSZ-660-supported CuNiAl-based catalysts in methanol, microwave heating,140 °C, 80 min	Lignin oil (69.8 wt%) containing lignin-derived dimers/trimers and phenolic monomers
Solvolytic depolymerization	Corn stover-derived Organosolv lignin in ethanol, 250 °C, 85 bar, 30 min	4-Vinyl guaiacol, 4-vinyl phenol as major monomers
Catalytic depolymerization	Raw lignin from corn straw, Ni–Fe–Mo$_2$C catalyst in methanol: water (4:1) 260 °C, 4 h	Liquefied lignin (89.56 wt%) containing phenolic monomers (35.53 wt%)
Catalytic hydrothermal liquefaction	Alkali lignin, biochar derived activated carbon-supported Ni–Co catalyst in ethanol, 280 °C, 15 min	Lignin oil (72.0 wt%) containing vanillin (34.8%) as major component
Catalytic hydrogenolysis	Organosolv poplar lignin, NiCu/C catalyst in ethanol/isopropanol (1:1), 1 MPa N$_2$, 270 °C, 4 h	Lignin oil containing phenolic monomers (63.4 wt%)
Catalytic hydrothermal liquefaction	Kraft lignin, Ni–Al/ MCM-41catalyst in ethanol, 280 °C, 1 h	Lignin oil (56.2 wt%) containing G-type phenolics
Catalytic hydroconversion	Kraft lignin, Re-Mo-supported zeolitic imidazolate framework nanocatalyst in 1, 4-dioxane and methanol, 300 °C 24 h	Monophenols and petroleum ether soluble biofuels
Reductive catalytic fractionation	Herbaceous biomass, MoO$_2$/C catalyst in methanol, 3 MPa H$_2$, 240 °C, 4 h	Removal of lignin from biomass with selective production of methyl coumarate and methyl ferulate

(continued)

Table 1.1 (continued)

Type of reaction	Types of lignin, catalysts, and reaction conditions	Products
Catalytic depolymerization	Pd-Al_2O_3-activated biochar catalyst in methanol, 3 MPa H_2, 240 °C, 3 h	30.4 wt% of C9 phenolic monomers yield with propyl guaiacol as major component
Catalytic transfer hydrogenolysis	Sugarcane bagasse, silicotungstic acid with carbon-supported Pd catalyst in isopropanol, 170 °C, 5 h	Lignin oil containing monophenols (34.91 wt%)

Adapted from Sethupathy et al. [7]. Reprinted with permission from Elsevier

References

1. D.S. Bajwa, G. Pourhashem, A.H. Ullah, S.G. Bajwa, A concise review of current lignin production, applications, products and their environmental impact. Ind. Crops Prod. **139**, 111526 (2019). https://doi.org/10.1016/j.indcrop.2019.11152
2. L. Dessbesell, M. Paleologou, M. Leitch, R. Pulkki, C. Xu, Global lignin supply overview and Kraft lignin potential as an alternative for petroleum-based polymers. Renew. Sustain. Energy Rev. **123**, 109768 (2020). https://doi.org/10.1016/j.rser.2020.109768
3. Statista, Market size of paper and pulp industry worldwide in 2021, with an estimated figure for 2022 and a forecast for 2029 (2022), https://www.statista.com/statistics/1073451/global-market-value-pulp-and-paper/. Accessed April 2024
4. Advanced Energy Technologies, Overview of bio energy technologies—cellulosic ethanol production (2024), https://aenert.com/technologies/renewable-energy/bio-energy/cellulosic-ethanol-production/. Accessed April 2024
5. Grand View Research, Lignin market size, share & trend analysis report by product (lingosulfonates, kraft lignin, organosolv lignin, others), by application, by region, and segment forecasts, 2024–2030 (2024), https://www.grandviewresearch.com/industry-analysis/lignin-market. Accessed April 2024
6. International Lignin Institute, What's lignin? (2024), https://www.ili-lignin.com/what-is-lignin.html. Accessed April 2024
7. S. Sethupathy, G.M. Morales, L. Gao, H. Wang, B. Yang, J. Jiang, J. Sun, D. Zhu, Lignin valorization: status, challenges and opportunities. Biores. Technol. **347**, 126696 (2022). https://doi.org/10.1016/j.biortech.2022.126696
8. ACS Green Chemistry Institute, 12 principles of green chemistry (2024), https://www.acs.org/greenchemistry/principles/12-principles-of-green-chemistry.html. Accessed April 2024
9. United Nations, The 17 goals (2024), https://sdgs.un.org/goals. Accessed February 2024
10. S. Vaz Jr., D.T. de Souza, Sustainable carbonaceous raw materials: a contribution to reduce the negative environmental impacts of chemical industry. Sustain. Chem. Environ. **5**, 100058 (2024). https://doi.org/10.1016/j.scenv.2024.100058

Chapter 2
The Biochemical Pathways in Plants

Abstract As the most abundant plant biomass when compared to the other classes, lignocellulosic biomass is formed by cellulose, hemicellulose, and lignin, which are the three components of the cell wall and the morphological structure of plants, been lignin a phenolic macromolecule. This chapter describes the physicochemical considerations of the lignin macromolecule, the metabolic pathways related to its biosynthesis, and differences between hardwoods and softwoods in order to understand the biochemistry aspects behind the natural production of lignin.

Keywords Lignocellulosic biomass · Chemical composition · Structure models · Metabolic pathways · Biosynthesis · Hardwood · Softwoods

2.1 The Physicochemical Considerations for the Lignin Macromolecule

Lignocellulosic biomass is the most abundant plant biomass when compared to the other classes (e.g., starchy, oleaginous, sugary) because it is formed by cellulose, hemicellulose, and lignin, which are the three components of the cell wall and the morphological structure of plants—cellulose (Fig. 2.1) and hemicellulose (Fig. 2.2) are polysaccharide molecules and lignin (Fig. 2.3) a phenolic macromolecule. Table 2.1 depicts the contents of lignocellulosic fractions for some crops and woods.

Despite the knowledge of the chemical precursors of the lignin structure [i.e., *p*-coumaryl alcohol or *p*-hydroxyphenil unit (H), coniferyl alcohol or guaiacyl unity (G), and sinapyl alcohol or syringyl unit (S)], aspects of tillage, plant species, and soil quality can affect directly the plant development and, consequently, the lignin formation and its physicochemical properties and characteristics.

In a general way, the lignin structure is generated from the repetition of the three monolignols as monomers (H, G, and S)—coniferyl and syringyl are the major monomeric unities—maintained by means a radically coupling and cross-linkages.

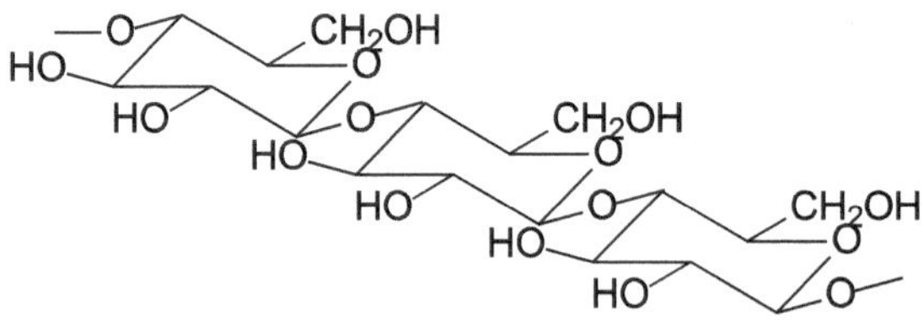

Fig. 2.1 Chemical structure of cellulose; the glucose units are linked by 1,4-β-D bond

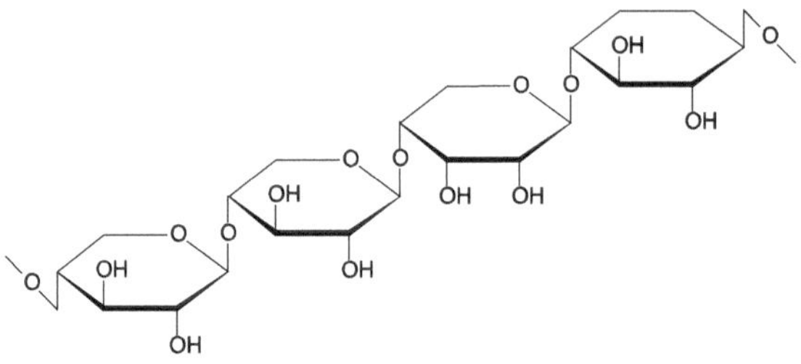

Fig. 2.2 Chemical structure of hemicellulose; the oligomeric units composed of D-glucose and pentoses (mainly D-xylose) are linked by means of a 1,4-β-D bond

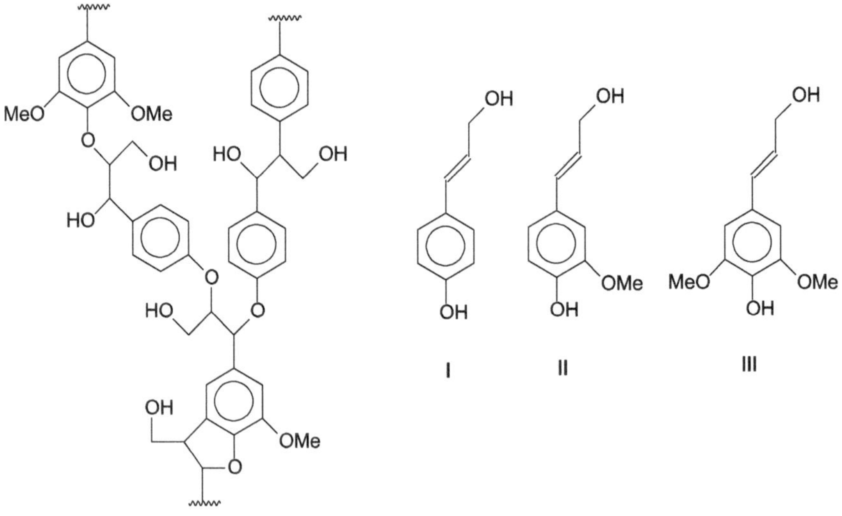

Fig. 2.3 Lignin structure (left) and its monolignol precursors (right): (I) *p*-coumaryl alcohol (the *p*-hydroxyphenyl unit, H); (II) coniferyl alcohol (the guaiacyl unity, G); and (III) sinapyl alcohol (the syringyl unit, S)

Table 2.1   Chemical composition of lignocellulosic biomasses, according Vassilev et al. [1]

Crops	%wt/wt cellulose	%wt/wt hemicellulose	%wt/wt lignin
Barley straw	48.6	29.7	21.7
Corn cobs	48.1	37.2	14.7
Grasses	34.2	44.7	21.1
Sugarcane bagasse	42.7	33.1	24.2
Rice husks	43.8	31.6	24.6
Wheat straw	44.5	33.2	22.3
Eucalyptus	52.7	15.4	31.9

Reproduced with permission from Elsevier

Regarding to the cross-linkages, they are promoted by these interunities linkages [2]:

- Phenylpropane β-aryl ether
- Resinol
- Phenylcoumaran
- Biphenyl
- Dibenzodioxocin
- Diaryl ether interunities linkages.

We can observe a high structural heterogeneity from the interunit linkages and functional groups attached to phenyl propane units. Interunit linkages are present in two major categories corresponding to ether bond linkages (e.g., β-O-4, α-O-4, 4-O-5) and carbon–carbon linkages (e.g., β-β, β-5, 5-5); and the relative abundance of each linkage in native lignin varies from plant to plant with the β-O-4 linkage being the most abundant [2].

Regarding those major functional groups present in lignin structures, we can observe:

- Hydroxyl (–OH)
- Methoxy (–CH$_3$)
- Carbonyl (–C=O)
- Carboxylic groups (–C=OOH).

And, as previously cited, the ratio of these groups presented in phenyl propane units is affected by the species genetic origin, isolation processes, and environmental issues [3].

However, as the advance of analytical techniques (to be seen in the Chap. 3) and the knowledge increasing, new monomers are discovered in order to refine and improve the definition of the lignin structure—as phenolics from beyond the monolignol biosynthetic pathway [4].

Figure 2.4 depicts relevant structure models for lignin from different species genetic origin.

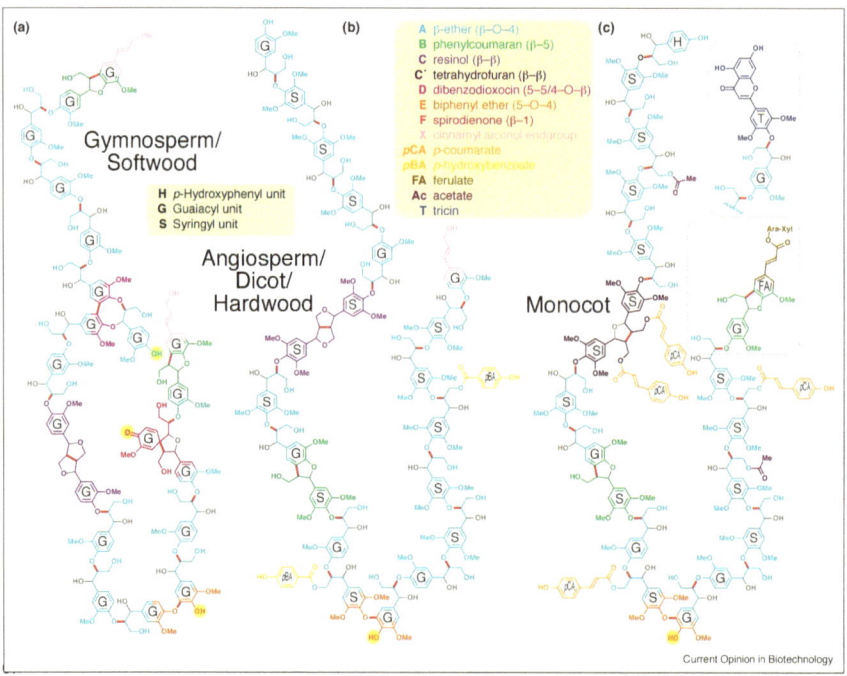

Fig. 2.4 Lignin model structures. Model lignin 20-mers are shown for: **a** a gymnosperm/softwood; **b** an angiosperm/dicot/hardwood; and **c** a monocot (commelinid). Authors considered limitations related to fit such data into a 20-mer. *Source* Ralph et al. [4]. Reprinted with permission from Elsevier

Softwoods and hardwoods (seen ahead) present a predominance of G unity, and G and S unities, respectively. On the other hand, monocotyledons present a predominance of S unity.

2.2   The Metabolic Pathways for the Lignin Biosynthesis

Metabolic pathways can define chemical species and reactions involved in the lignin synthesis (or biosysnthesis) in plants. Generally, they comprise organic compounds—highlighting those with aromatic groups—with oxygen and nitrogen atoms and aliphatic terminations; and reactions are catalyzed by several enzymes.

The phenylpropanoid pathway serves the biosynthesis of many metabolic classes. The actual metabolic route in a given cell depends on how the pathway is regulated [5].

Figure 2.5 depicts metabolic routes related to lignin biosynthesis in plants.

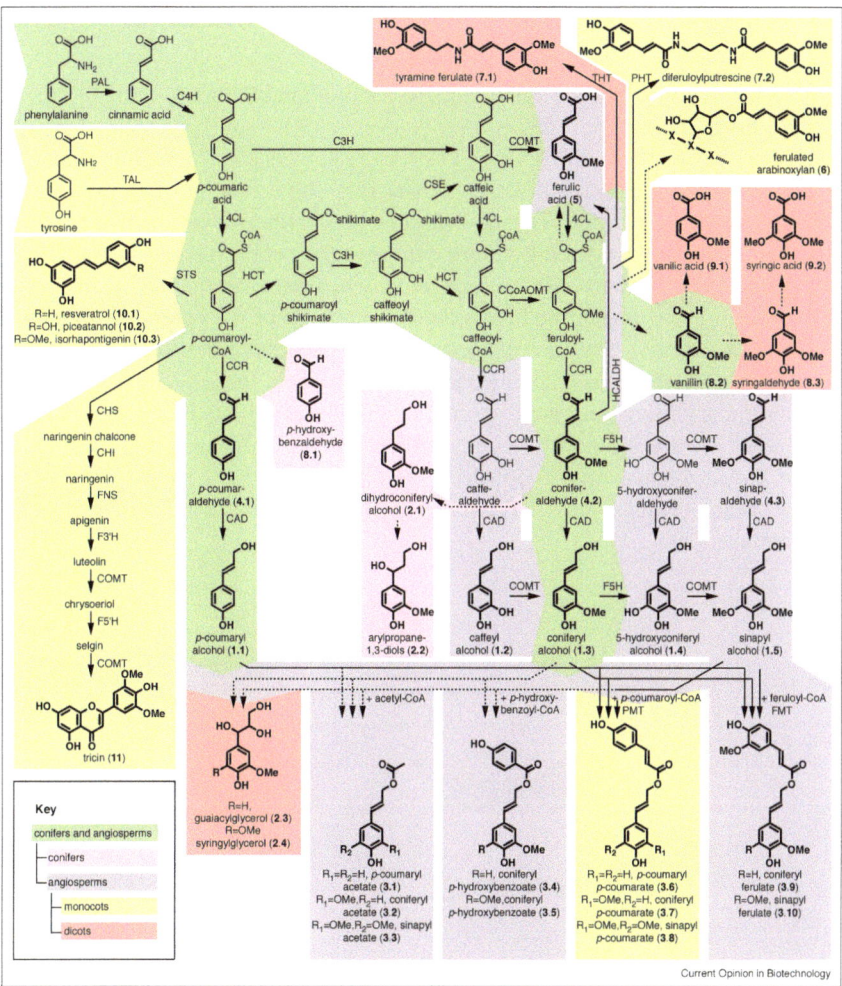

Fig. 2.5 Metabolic pathways leading to the lignin macromolecules and its monomers. Lignin monomers are shown in bold and are followed by a number indicating the respective metabolic class and the arbitrary ranking within this class. Some monomers occur in the lignin of both conifers and angiosperms, but others have, to date, primarily been found in the lignin of one of the classes—conifers, angiosperms, dicots, and/or monocots (indicated by a different color). Solid arrows represent enzymatic steps evidenced by at least in vitro activities, dashed arrows represent suggested pathways. *PAL* phenylalanine ammonia-lyase; *TAL* tyrosine ammonia-lyase; *C4H* cinnamate 4-hydroxylase; *4CL* 4-coumarate:CoA ligase; *HCT* p-hydroxycinnamoyl-CoA:quinate/shikimate p-hydroxycinnamoyltransferase; *C3H* p-coumarate 3-hydroxylase; *CSE* caffeoyl shikimate esterase; *CCoAOMT* caffeoyl-CoA O-methyltransferase; *CCR* cinnamoyl-CoA reductase; *F5H* ferulate 5-hydroxylase; *COMT* caffeic acid O-methyltransferase; *CAD* cinnamyl alcohol dehydrogenase; *LDH* hydroxycinnamaldehyde dehydrogenase; *PER* peroxidase; *FMT* feruloyl-CoA monolignol transferase; *PMT* p-coumaroyl-CoA monolignol transferase; *CHS* chalcone synthase; *CHI* chalcone isomerase; *FNS* flavone synthase; *F3′H* flavonoid 3′-hydroxylase; *F5′H* flavonoid 5′-hydroxylase; *THT* hydroxycinnamoyl-CoA:tyramine N-hydroxycinnamoyltransferase; *PHT* hydroxycinnamoyl-CoA:putrescine hydroxycinnamoyl transferase; *STS* stilbene synthase. *Source* Vanholme et al. [5]. Reprinted with permission from Elsevier

2.2.1   Differences Related to Lignocellulosic Biomasses

Firstly, it is worth introduce some basic issues of plant biology.

We have main plant classes: *angiosperms* and *gymnosperms* (Fig. 2.6). Angiosperms are plants characterized by having flowers and fruits. This plant group is the one with the greatest diversity of species, with a total of more than 450,000 different species estimated. They are vascular plants (they have conducting vessels) with seeds whose most striking feature is this presence of flowers and fruits. The term angiosperm comes from the Greek *angeion*, which means urn, and *sperma*, which means seed. Fruits are formed from the development of the flower ovary after the fertilization process. They are important for the success of this group of plants, as they act to protect the seed and help in the dispersion of these structures. Currently, it is customary to classify angiosperms into *monocotyledons* (or monocots), *eudicotyledons* (or dicotyledons or eudicots/dicots), *magnolides* and a group of plants called "basal angiosperms". However, the largest groups of angiosperms are the monocots and eudicots/dicots. These two groups have some characteristics that help differentiate them.

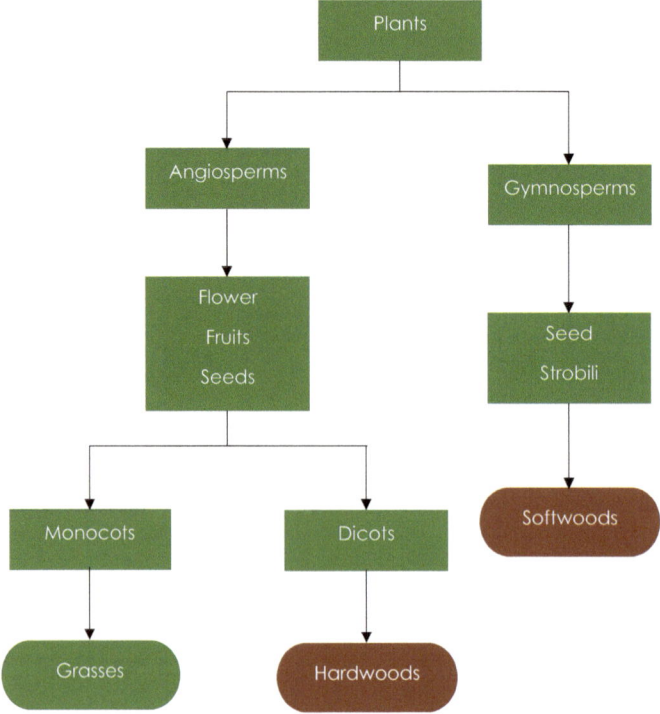

Fig. 2.6 Plants classification addressed to generate hardwoods and softwoods

Gymnosperms are part of the group of vascular plants with seeds, as are angiosperms. However, in gymnosperms, the occurrence of seeds without the presence of fruits surrounding them is observed. In gymnosperms, the presence of flowers is not observed, and, in some cases, the strobilus is mistakenly called that way. Strobili, also called cones, are, in reality, reproductive structures that have modified leaves capable of producing spores. In gymnosperms, we find strobili capable of producing pollen and strobili capable of producing ovules.

Thus, the main difference between gymnosperms and angiosperms is the fact that gymnosperms produce seeds, but they are naked seeds, that is, without fruit involving this structure. In addition to the absence of fruits, the presence of flowers is also not observed in gymnosperms. Examples of gymnosperms are pine trees.

The most common woods for industrial processing are divided into *hardwoods* and *softwoods*, each one with their lignin characteristic and typical content.

Hardwoods come from dicot trees (e.g., eucalyptus from family *Myrtaceae*; Fig. 2.7), which reproduce through flowers. They are commonly found in broad-leaved temperate and tropical forests. While temperate and boreal hardwoods are mostly deciduous, tropical ones may shed leaves in response to seasonal or sporadic droughts.

Hardwoods—come from gymnosperm trees (e.g., pinus from *Pinaceae* family; Fig. 2.8)—have broad leaves (as opposed to needle-like leaves in softwoods). Many temperate hardwoods lose their leaves every autumn and remain dormant in winter. Some tropical hardwoods may retain their leaves even during dry periods.

Fig. 2.7 Eucalyptus tree plantation at Ooty, India. *Source* Shutterstock

Fig. 2.8 Pine trees in mist on the island of São Miguel, Azores, Portugal. *Source* Shutterstock

Hardwoods exhibit annual growth rings, although some tropical species lack them—their growth is often slower compared to softwoods. The most significant feature distinguishing hardwoods from softwoods is the presence of *pores* or *vessels* in their wood. These pores or vessels vary in size, shape, and structure, including spiral thickenings. While hardwoods are generally harder than softwoods, there are exceptions.

Regarding the chemical composition, we can observe cellulose, hemicellulose and lignin in the cell wall for structural purposes for both classes. Phenolic compounds, such as stilbenes, lignans, tannins, and flavonoids, are abundant in hardwoods. However, hardwood lignin differs from that in softwoods lignin can present higher total OH, lower β-O-4′, and higher molecular weight than hardwood's [6].

As commented before, softwood refers to wood derived from gymnosperm trees, such as pines and spruces. These trees have needle-like or scale-like leaves. Unlike hardwoods, softwoods lack resin canals and pores (Fig. 2.9).

Softwoods are not necessarily softer than hardwoods. In fact, some softwoods, like longleaf pine, Douglas fir (family *Pinaceae*), and yew (family *Taxaceae*), are much harder than several hardwoods.

According to this functionality in the plant structure, lignin act as a "glue" that holds together the cellular-wall of lignocellulosic biomass, giving robustness and protection (Fig. 2.10).

The differences related to chemical and structural characterization of hardwood and softwood obtained by the LignoForce™ process were studied by [6]. The results are depicted in Table 2.2.

Fig. 2.9 SEM images showing the presence of pores in hardwoods (oak, top) and absence in softwoods (pine, bottom). *Source* Wikipedia (en.wikipedia.org/wiki/Softwood)

We can observe from Table 2.21 an experimental evidence that lignin can vary from hardwood to softwood for monomers, interunits, end-groups, and others (methoxyl and xylose), and this variation can be associated with metabolic routes of biosynthesis for each wood type. Furthermore, as noted by Gellerstedt and Henriksson [8], softwood lignin consists exclusively of coniferyl alcohol (G unity), while hardwood lignin consists mainly of coniferyl alcohol and sinapyl alcohol (S unity)—it can be observed in Fig. 2.4.

Even though they are not the main source of technical lignin, grasses (Fig. 2.11) are family of plants (*Poaceae*). Its representatives consist of flowering, monocotyledonous plants (Fig. 2.6), and it is a prominent family among the families economically relevant to humans.

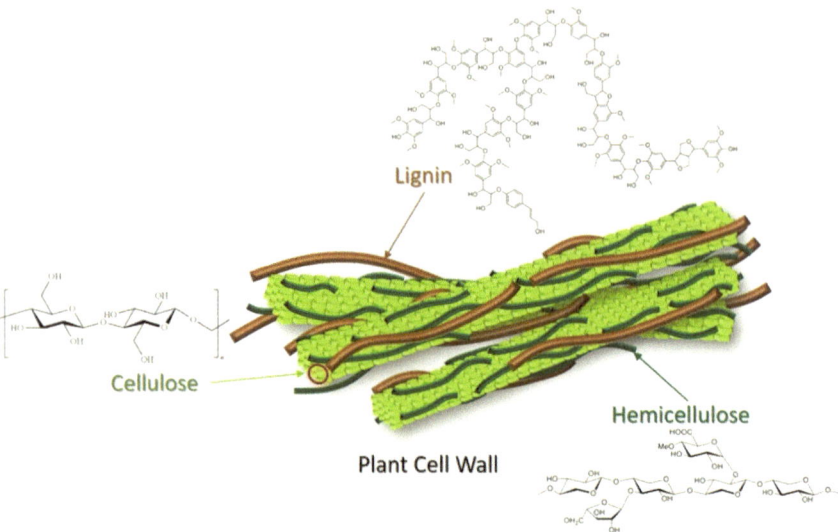

Fig. 2.10 Structure of lignocellulosic biomass. *Source* Jensen et al. [7]. Reprinted with permission from Springer Nature

Some representatives of *Poaceae* are corn (*Zea mays*), wheat (*Triticum aestivum*), rice (*Oryza sativa*), oat (*Avena sativa*), sugarcane (*Saccharum officinarum*), and barley (*Hordeum vulgare*) are just a few examples. Some are cultivated for construction, like bamboo (*Phyllostachys aurea*) in Asia. Others are grown for food and feed, like corn. And others are simply cultivated to meet the demand for the production of delicacies in global cuisine, such as rice, oat and barley, and many others.

Some grasses commonly used to obtain lignin are sugarcane (*S. officinarum*), elephant grass (*Pennisetum purpureum* Schumac), and sweet grass (*Hierochloe odorata* or *Anthoxanthum nitens*). Table 2.1 shows the chemical composition of some grasses.

2.3 Conclusions

Lignin is one of the three components of the lignocellulosic biomass structure and responsible for maintaining the cellular-wall and the plant morphology. Its biosynthesis is promoted by several metabolic pathways suffering influences from genetic and environmental factors. Moreover, the lignin content can vary in lignocellulosic biomasses (e.g., softwoods and hardwoods).

Table 2.2 Quantification of major substructures in LFHL and LFSL, with their relative abundance expressed in relation to 100 lignin monomers

Lignin structural quantification per 100 C9 units	LFHL	LFSL
Lignin monomers		
Syringyl (S)	66.3	nd
Guaiacyl (G)	33.7	96.2
p-hydroxyphenyl (H)	nd	3.8
S/G ratio	1.97	nd
H/G ratio	nd	0.03
Lignin interunits		
Aryl ether (β-O-4′)	6.2	5.8
Phenylcoumaran (β-5′)	0.2	1.7
Resinol (β-β′)	3.6	2.1
Epiresinol (β-β′)	1.6	0.3
Secoisolariciresinol (β-β′)	nd	1.6
Stilbene (β-1′)	1.0	0.5
Stilbene (β-5′)	2.1	8.1
Lignin end-groups		
Cinnamyl alcohol	nd	1.4
Cinnamaldehyde	0.8	2.0
p-Hydroxybenzoate	0.6	nd
Ferulate	1.2	2.1
Others		
Methoxyl[a]	185	117
Xylose	0.4	0.6

Source Suota et al. [6]. Reprinted with permission from Elsevier

[a]The quantification of methoxyl groups by HSQC-NMR might have been overestimated in LFSL due to some signal overlap; *LFHL* LignoForce™ hardwood lignin; *LFSL* LignoForce™ softwood lignin; *nd* not detected

Fig. 2.11 Examples of grasses for lignin obtention, where **a** sugarcane (*S. officinarum*); and **b** elephant grass (*P. purpureum* Schumac). *Source* Shutterstock

References

1. S.V. Vassilev, D. Baxter, L.K. Andersen, C.G. Vassileva, T.J. Morgan, An overview of the organic and inorganic phase composition of biomass. Fuel **94**, 1–33 (2012). https://doi.org/10.1016/j.fuel.2011.09.030
2. S.C.D. Eswaran, S. Subramaniam, U. Sanyal, R. Rallo, X. Zhang, Molecular structural dataset of lignin macromolecule elucidating experimental structural compositions. Sci. Data **9**, 647 (2022). https://doi.org/10.1038/s41597-022-01709-4
3. H.A. Ariyanta, F.P. Sari, A. Sohail, W.K. Restu, M. Septiyanti, N. Aryana, W. Fatriasari, A. Kumar, Current roles of lignin for the agroindustry: applications, challenges, and opportunities. Int. J. Biol. Macromol. **240**, 124523 (2023). https://doi.org/10.1016/j.ijbiomac.2023.124523
4. J. Ralph, C. Lapierre, W. Boerjan, Lignin structure and its engineering. Curr. Opin. Biotechnol. **56**, 240–249 (2019). https://doi.org/10.1016/j.copbio.2019.02.019
5. R. Vanholme, B. De Meester, J. Ralph, W. Boerjan, Lignin biosynthesis and its integration into metabolism. Curr. Opin. Biotechnol. **56**, 230–239 (2019). https://doi.org/10.1016/j.copbio.2019.02.018
6. M.J. Suota, T.A. da Silva, S.F. Zawadzki, G.L. Sassaki, F.A. Hansel, M. Paleologou, L.P. Ramos, Chemical and structural characterization of hardwood and softwood LignoForce™ lignins. Ind. Crops Prod. **173**, 114138 (2021). https://doi.org/10.1016/j.indcrop.2021.114138

7. C.U. Jensen, J.K. Rodriguez-Guerrero, S. Karatzos, G. Olofson, S.B. Iversen, Fundamentals of hydrofaction™: renewable crude oil from woody biomass. Biomass Conversion Biorefinery **7**, 495–509 (2017). https://doi.org/10.1007/s13399-017-0248-8

8. G. Gellerstedt, G. Henriksson, in *Chapter 9—Lignins: Major Sources, Structure and Properties*, ed. by M.N. Belgacem, A. Gandini. Monomers, Polymers and Composites from Renewable Resources (Elsevier, 2008), pp. 201–224. https://doi.org/10.1016/B978-0-08-045316-3.00009-0

Chapter 3
Advanced Analytical Techniques for Lignin

Abstract The understanding of the physicochemical properties and the chemical composition of the lignin macromolecule is of fundamental importance for its technological usage. However, it will depend on the application of a large set of analytical techniques for these purposes. This chapter treats about the description of the most used advanced analytical chemistry, a critical application of these analytical techniques for Kraft lignin as a case study, and new analytical approaches in order to generate analytical information for the macromolecule usages.

Keywords Spectroscopic techniques · Chromatographic techniques · Thermal techniques · Microscopy · Surface techniques

3.1 Description of Most Used Advanced Analytical Techniques Available for Lignin

Lignocellulosic materials, or lignocellulosic biomass, are formed by hard and fibrous structures consisting mainly of the polysaccharides cellulose and hemicellulose (around 70% of the dry mass), interspersed with a phenolic macromolecule, lignin (around 30% of the dry mass), to which it is linked by covalent and hydrogen bonds. In smaller proportions, and depending on the origin of the species, resins, fatty acids, phenols, tannins, nitrogenous compounds, and mineral salts can also be found, besides calcium, potassium, and magnesium [1].

As previously introduced in Chap. 2, lignin is a component of the plant cell-wall, normally considered recalcitrant and also an inhibitor of the digestibility of plants, accentuating its action as the plant ages [2]. From a functional point of view, lignins facilitate water transport, provide resistance to cell walls, and prevent degradation

The original version of the chapter has been revised. A correction to this chapter can be found at https://doi.org/10.1007/978-3-031-75511-8_9

© The Author(s), under exclusive license to Springer Nature Switzerland AG 2024, corrected publication 2025
S. Vaz Jr., *The Lignin Macromolecule*,
SpringerBriefs in Applied Sciences and Technology,
https://doi.org/10.1007/978-3-031-75511-8_3

of wall polysaccharides, acting as an important line of defense against pathogens, insects, and other herbivores [3].

Therefore, understanding the physicochemical properties and the chemical composition of the lignin macromolecule is of fundamental importance for its technological usage. However, it will depend on the application of a large set of analytical techniques for these purposes.

Due to its chemical structure, lignin is insoluble in most organic solvents. Therefore, isolation is difficult. However, when their separation is achieved, their original molecular structure is generally compromised. Although it is not possible to extract lignin without degrading it, it is estimated that its molecular mass may range from 1000 to 20,000 g mol^{-1}; and when isolated, they have a dark color, being relatively stable in mineral acid solutions and soluble in hot aqueous bases [4].

El Khaldi-Hansen et al. [5] suggest several analytical techniques and analytical methods for analyzing lignin from temperate softwoods, which can be adapted to hardwood lignin from tropical climates and more commonly used in Kraft processing in the cellulose and paper industry. Such techniques and methods include:

 (i)   Elucidation of the molecular structure;
(ii)   And chemical composition and determination of physical–chemical properties.

Therefore, the typical analytical techniques to be used and the information generated from them comprise:

• Thermogravimetric analysis (TGA): to observe the thermal decomposition (mass loss due to the temperature increasing) of the sample.
• Surface area analysis of the sample by means the Brunauer–Emmett–Teller (BET) method: to determine a physicochemical property.
• Differential scanning calorimetry (DSC): to observe the glass transition temperature of the sample in order to determine physicochemical properties.
• Size exclusion liquid chromatography with enhanced light scattering detector (SEC-ELSD): to observe the size distribution of the molecule in order to determine physicochemical properties.
• Gas chromatography with a pyrolysis probe coupled to a mass spectrometry detector (Py-GC-MS): for structural elucidation.
• Infrared absorption spectroscopy with Fourier transform (FTIR): for structural elucidation.
• Scanning electron microscopy combined with the identification of the chemical elements present in the sample by X-ray spectrometry (SEM-EDS): to observe the morphology of the sample surface in order to determine physicochemical properties and to determine its chemical composition.
• Solid-state ^{13}C nuclear magnetic resonance (^{13}C-NMR), with magic angle: for structural elucidation. It can be complemented with other nuclei (e.g., ^{31}P-NMR).

Thus, this set of analytical techniques and their analytical methods is paramount to evaluate the use of lignin as a carbonaceous raw material for several products (e.g., chemicals and materials).

3.2   A Critical Application of Analytical Techniques as a Case Study

The analytical techniques introduced in the item 3.1 will be treated in this item in order to clarify their practical application considering a sample of hardwood lignin from *Eucalyptus grandis* × *Eucalyptus urophylla* hybrid plant. And the operating conditions of each technique are briefly described as follow.

3.2.1   Thermogravimetric Analysis

The thermogravimetric curve and its derivative as a function of temperature can be obtained on a TGA Q500 thermal analyzer (TA Instruments), using a heating rate of $20\,°C\,min^{-1}$ with an alumina sample support. The specimen (sample) is heated from room temperature (near 25 °C) to 550 °C, under a nitrogen atmosphere, at a flow rate of $50\,mL\,min^{-1}$. After reaching 550 °C, the nitrogen is replaced by oxygen, at the same flow rate, with heating up to 950 °C. The analytical method will generate a result as depicted in Fig. 3.1.

A block diagram for the thermal analyzer is shown in Fig. 3.2.

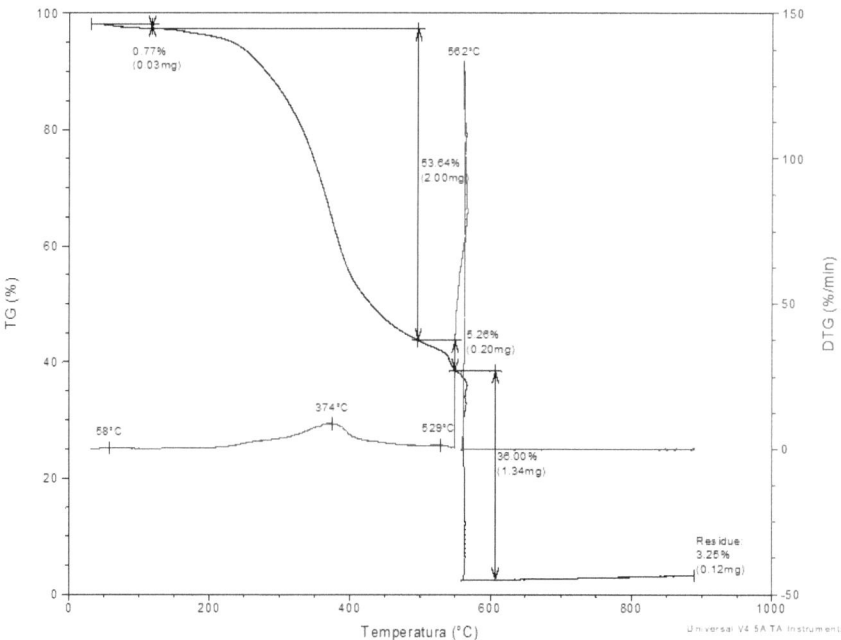

Fig. 3.1 A thermogram of eucalyptus Kraft lignin. First derivative curve in blue. Adapted from Vaz Júnior et al. (2020)

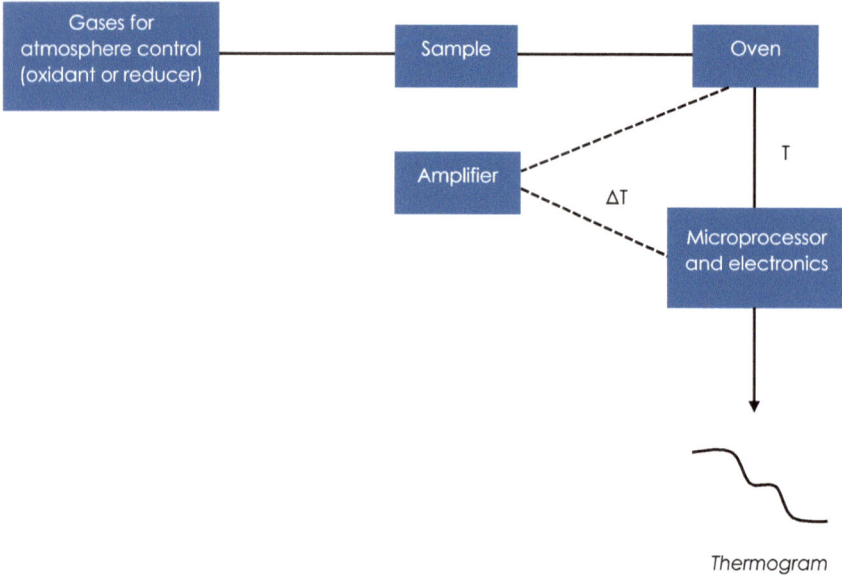

Thermogram

Fig. 3.2 Simplified block diagram of a TGA instrument

According to Kim et al. (2014), the thermal degradation of wood lignin occurs over a wide temperature range, starting at around 154 °C and extending to very high temperatures. In this way, it was possible to observe this same thermal behavior in the analyzed sample. It should be noted that, in the range of approximately 200–300 °C, the degradation of aliphatic alcohols, acids and esters occurs [6], with a peak being observed at 374 °C (first derivative of the thermogravimetric curve), which is indicative of the presence of such groups. The same derivative provided a very sharp peak at 562 °C, probably due to the decomposition of aromatic rings (Kim et al. 2014).

3.2.2 Superficial Area Analysis by BET Method

The lignin sample is subject to a previous vacuum temperature treatment to eliminate contamination acquired by exposure to the ambient atmosphere. After this treatment, the sample is cooled with liquid nitrogen in vacuum until the equilibrium temperature (− 200 °C). The adsorbate (high-purity nitrogen) is inserted in a controlled manner into the test chamber with the adsorption being measured until the equilibrium pressure is reached. From this equilibrium pressure, the amount of adsorbate adsorbed to the surface of the sample was calculated and, therefore, the surface area obtained. The test can be carried out using a FlowSorb II 2300 equipment (Micromeritics), with data

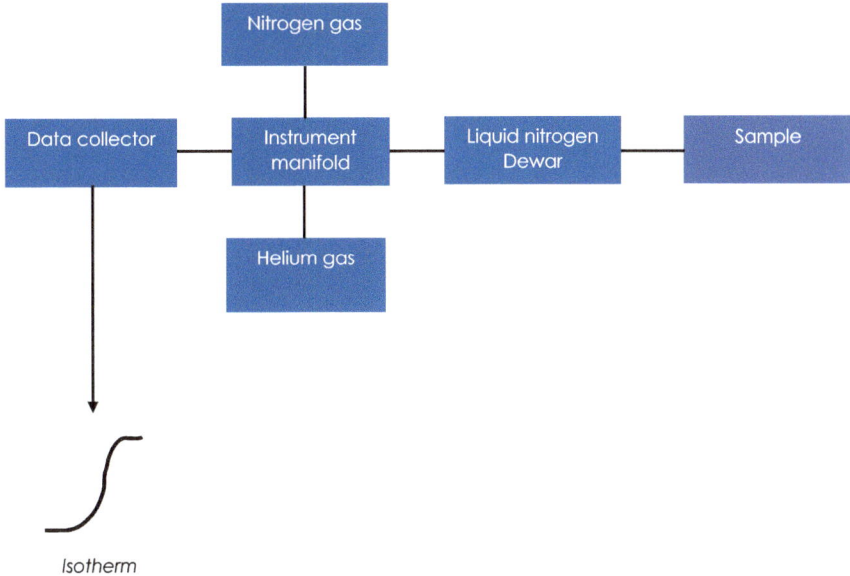

Fig. 3.3 Simplified block diagram for a BET instrument

interpretation using the Brunauer–Emmett–Teller (BET) theory of gas adsorption in multi-layers [7].

Literature values for lignin are in the order of 1.2 m^2 g^{-1} [8]. Therefore, it is expected that the greater the surface area of the sample, the greater its adsorption capacity, due to a greater availability of intermolecular interaction sites, promoted by –OH, alkyl, and aryl groups.

Figure 3.3 presents a block diagram for a BET analyzer.

3.2.3 ^{13}C-NMR Analysis in the Solid State

An Avance 400 MHz NMR spectrometer (Brüker) can be used to characterize the carbons according to their chemical environment, operating at a resonance frequency of 100.58 MHz, cross-polarization spectral band of 50 kHz, contact time of 1 ms, 500 ms repetition time, 0.0128 ms acquisition time, 0–230 ppm scan. The sample is inserted into a 5 mm zirconia rotor, with a magic angle of 6.4 kHz and a ^{1}H channel ramp of 110–60% (in kHz), according to Novotny et al. [9]. The analytical method will generate a result as depicted in Fig. 3.4. The spectrometer block diagram is depicted in Fig. 3.5.

The following assignments to groups, that depends on the chemical shift, can be made:

- 0–45 ppm: unsubstituted aliphatic C due to methyl-terminal groups.

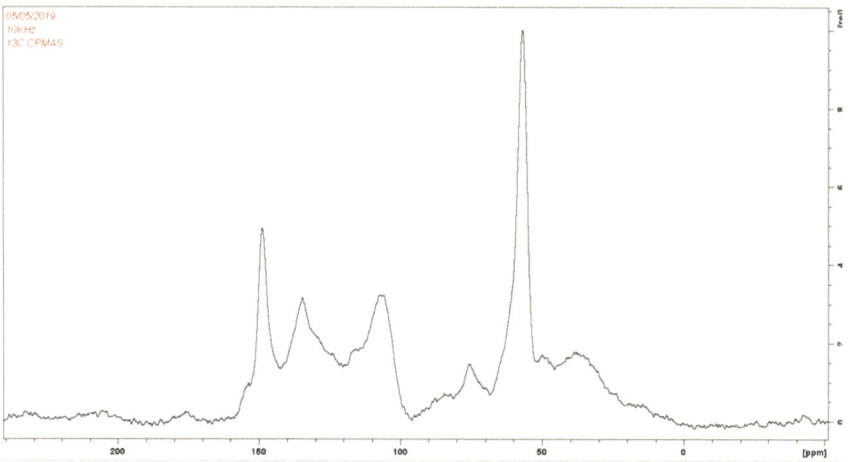

Fig. 3.4 Spectrum of the eucalyptus Kraft lignin obtained by the ^{13}C-NMR technique in the solid state with crossed polarization and magic angle modes. Adapted from Vaz Júnior et al. (2020)

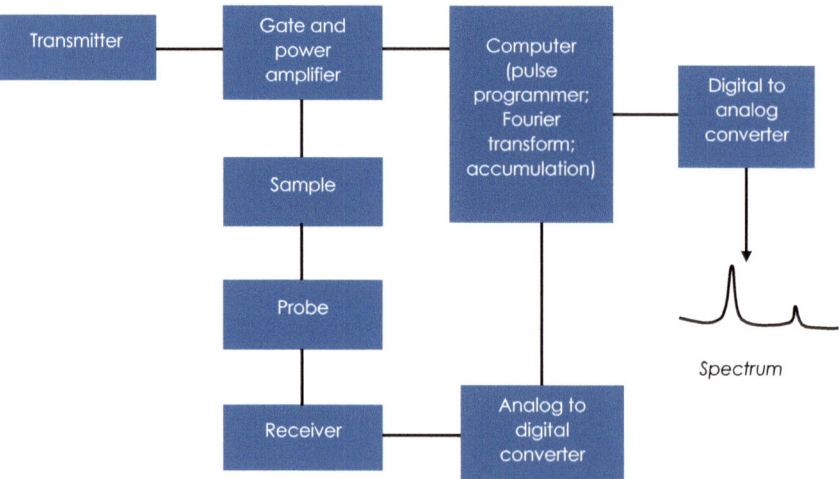

Fig. 3.5 Simplified block diagram for a pulsed Fourier transform-NMR instrument

- 45–65 ppm: C associated with *N*-alkyl, as in proteins, and methoxylic C.
- 60–110 ppm: C associated with aliphatic O.
- 110–140 ppm: unsubstituted and alkyl-substituted aromatic C.
- 110–160 ppm: total aromatic C related to unsubstituted, alkyl-substituted, and phenolic groups.
- 140–160 ppm: phenolic C.

3.2.4 Calorimetric Analysis by DSC

The DSC curves of the sample can be obtained using a capped aluminum sample holder. The sample is heated from room temperature (23 °C) to 250 °C at a heating rate of 20 °C min^{-1} and then cooled to 23 °C at 20 °C min^{-1} and heated again to 250 °C at 20 °C min^{-1}. The analyses can be carried out in a high-purity nitrogen atmosphere, at a flow rate of 50 mL min^{-1} in a Polyma DSC 214 equipment (Netzsch). The analytical method will generate a result as depicted in Fig. 3.6. The block diagram of the equipment is depicted in Fig. 3.7

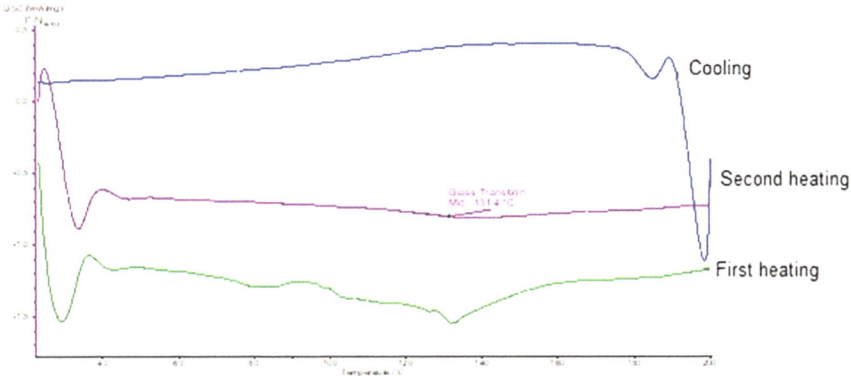

Fig. 3.6 DSC curves obtained for eucalyptus Kraft lignin. Adapted from Vaz Júnior et al. (2020)

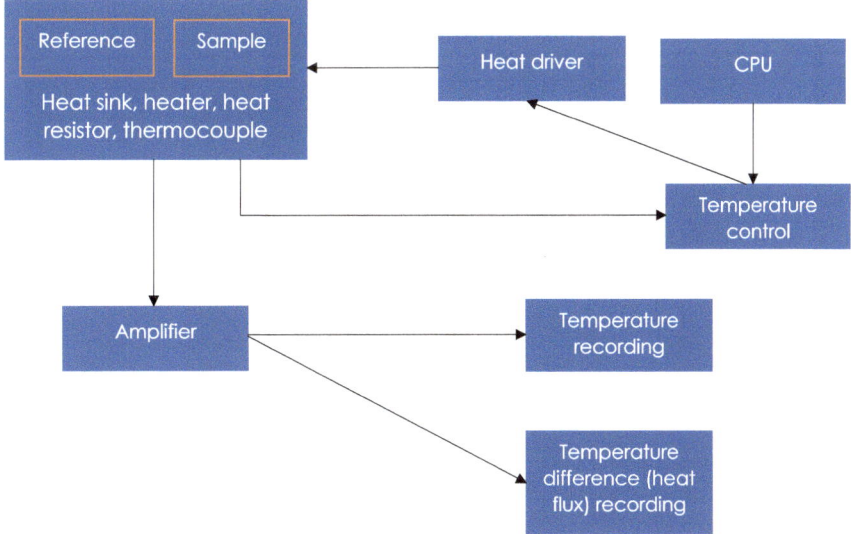

Fig. 3.7 Simplified block diagram of a DSC analyzer

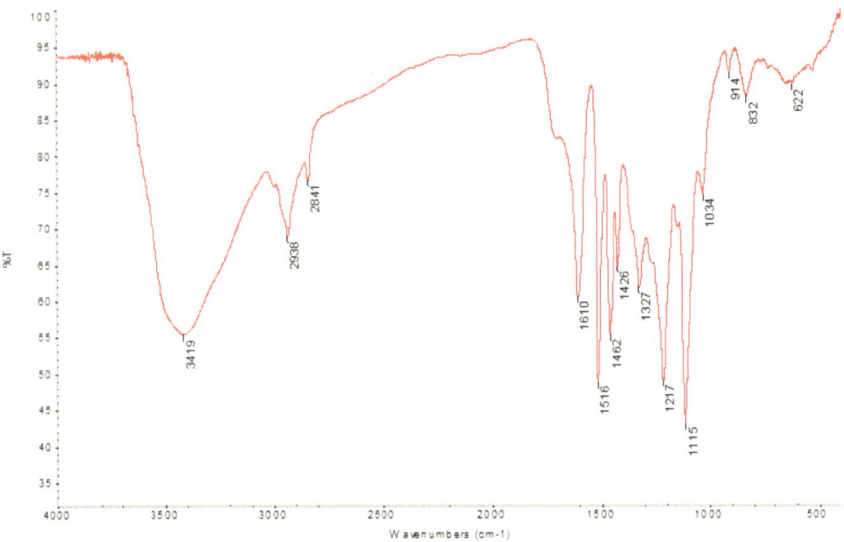

Fig. 3.8 Absorption spectrum in the mid-infrared region of eucalyptus Kraft lignin. Adapted from Vaz Júnior et al. (2020)

We can observe in Fig. 3.6 that the second heating of the sample provided a T_g value (glass transition temperature) of 131.4 °C, a temperature typical of an amorphous polymer, such as lignin [10]. It is worth noting that T_g is the main thermal transition of interest to the study.

3.2.5   Infrared Spectroscopic Analysis by FTIR

The IR spectra can be obtained with 32 repetitions at a reading range of 4000–400 cm^{-1} and a resolution of 4 cm^{-1}. The sample is prepared in anhydrous KBr tablet at a mass ratio of 99:1%wt/wt. A Nexus 4700 FTIR spectrometer (Thermo Nicolet) can be used. The analytical method will generate a result as depicted in Fig. 3.8. And the block diagram of the spectrometer is depicted in Fig. 3.9.

Based on the main bands observed in the spectrum of Fig. 3.8, we can perform their assignments as demonstrated in Table 3.1 with typical bands observed for lignins.

3.2.6   Microscopic Analysis by SEM-EDS

The scanning electron microscopy (SEM) is used for surface imaging. It is based on the scanning of the surface of the sample by a narrow beam of 10 nm of primary electrons with energy in the order of 10 keV, which leads to the construction of the

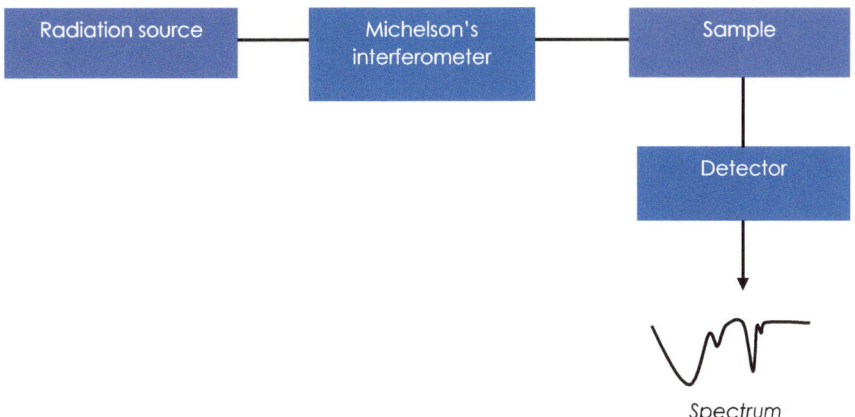

Fig. 3.9 Simplified block diagram of a MIR (mid-infrared) spectrometer with Fourier transform (FTIR)

Table 3.1 Assignment of absorption bands in the mid-infrared region of eucalyptus Kraft lignin. Adapted from Vaz Júnior et al. (2020).

Wavenumber (cm^{-1})	Band intensity	Assignment
3419	Strong	H–O
2938	Medium	OCH$_3$
2841	Weak	CH$_3$ and CH$_2$
1610	Medium	C=O
1516	Medium	Benzene ring
1462	Medium	Benzene ring
1426	Weak	Alkane
1217	Medium	C–O
1115	Medium	C–O
1034	Weak	C–O
914	Weak	C=C–H
832	Medium	C=C–H

image. Probably, this is the most widely used microscopic technique for examining polymers and materials in general. A relevant SEM characteristic is that it can be coupled to an EDS (energy-dispersive spectroscopy)—a X-ray-type spectroscopy—to produce quantitative information about the elemental constitution of the material surface. We can use a PhiZAF spectrometer (EDAX) coupled to the electron microscope Inspec (FEI), operating in BSED (backscattering electron diffraction) and SE (secondary electron) modes to analyze powder lignin samples.

The sample, after preparation, is placed on a metal support (stub) and fixed with double-sided C tape to improve the conductivity of the material. The analytical method will generate a result as depicted in Fig. 3.10, and the elementary composition

information in Table 3.2. And the block diagram of the microscope is depicted in
Fig. 3.11.

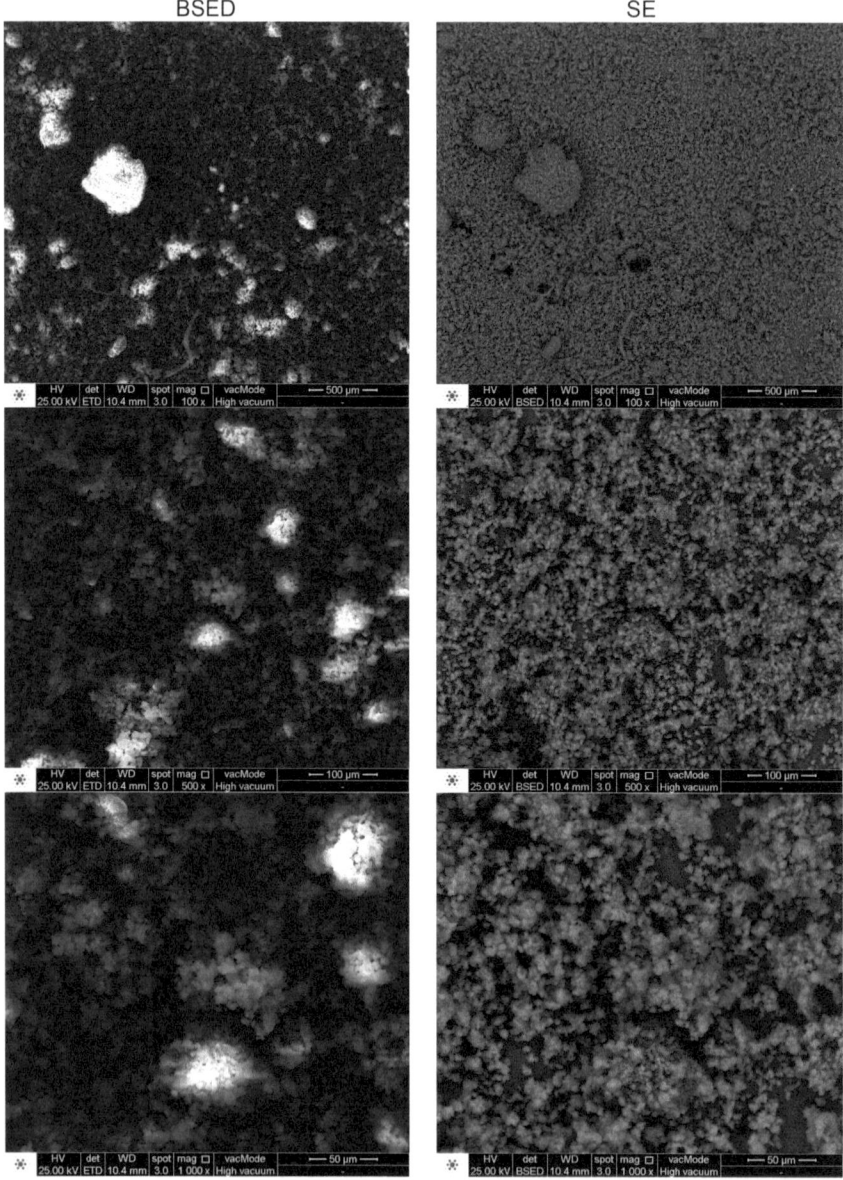

Fig. 3.10 Micrographs of eucalyptus Kraft lignin. BSED (backscattering electron diffraction) and
SE (secondary electron) modes. Adapted from Vaz Júnior et al. (2020)

Table 3.2 Elemental composition of eucalyptus Kraft lignin. Adapted from Vaz Júnior et al. (2020)

Element	%wt/wt
C	65.93
O	31.17
Na	0.36
Al	0.27
S	2.26

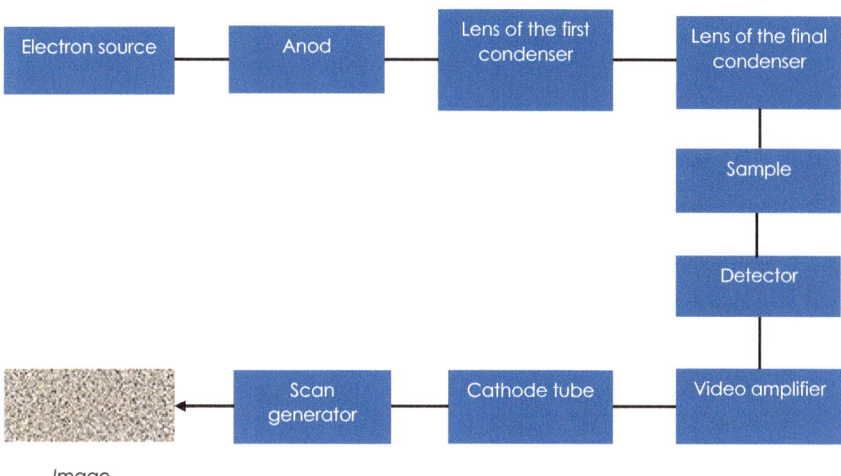

Image

Fig. 3.11 Simplified block diagram for a SEM-EDS microscope

Through micrographs (Fig. 3.10), it is possible to observe the morphological characteristics of lignin as a typical porous polymeric material, demonstrating uniformity and homogeneity of the sample surface, which facilitates, for instance, processes of slow release of organic compounds incorporated into the lignin molecule [11].

As commented before, the compositional analysis by EDS provides the mass percentages presented in Table 3.2.

As expected, the highest element content present is carbon (65.93%wt/wt), followed by oxygen (31.17%wt/wt). It is worth noting that the lignin molecule is rich in oxygenated groups, such as hydroxyls and ethers. Such values are in line with those obtained by ^{13}C-NMR, FTIR, and Py-GC-MS (seen ahead).

The presence of Na, Al, and S is probably due to inorganic residues from the Kraft processing of the eucalyptus wood, specially S, as introduced in Chap. 1 (item 1.2).

3.2.7 Chromatographic Analysis by Py-GC-MS

A small wad of quartz wool is inserted into the capillary, also made of quartz, until approximately half of the latter. The sample is inserted through the opposite opening and weighed immediately afterward (mass of 0.60 mg). Soon after, another wad of quartz wool is inserted, so that the sample is positioned in the middle of the capillary between the two wads. The chromatographic conditions are as follows: (i) split injector: 100:1, 280 °C; (ii) pressure flow control: pressure of 200 kPa; 3 mL min^{-1} for purge; (iii) temperature programming: 40 °C for 8 min, heating at 3 °C min^{-1} to 260 °C and isotherm at 260 °C for 40 min. We can use a Zebron ZB1701 column (Phenomenex) of 60 m × 0.25 mm × 0.25 μm. A triple quadrupole detector; an ion source temperature of 240 °C; interface temperature of 260 °C; acquisition mode of Q3 scan, m/z 40–400. For integration, the 30 peaks with the largest area are considered, one of which should be discarded because it could be a component originating from column bleeding. Then, pyrolysis is carried out under the following conditions: (i) probe: initial temperature of 200 °C for 3 min, ramp 20 °C ms^{-1}; (ii) final temperature of 60 °C for 20 s. Interface: stand by at 45 °C, initial temperature of 200 °C, ramp of 100 °C min^{-1}, and final temperature of 250 °C for 5 min. Transfer line at 300 °C, valve oven at 280 °C. Purge gas: industrial nitrogen. The equipment used can be a gas chromatographer GCMS-TQ8040 Shimadzu with a probe for pyrolysis and mass spectrometry detector coupled to it.

The analytical method will generate a result as depicted in Fig. 3.12 and the block diagram for the equipment is depicted in Fig. 3.13.

In Table 3.3 are identified the main compounds resulting from the pyrolytic degradation of the sample. The monomeric unit present in the largest quantity (68%) was the syringyl unit (S), or sinapyl alcohol, followed by the guaiacyl unit (G), 23%, and p-hydroxyphenyl (H), 5%. This result is typical of lignin extracted from E. grandis and E. urophylla [12, 13]. Furthermore, and as expected, the majority of the identified compounds are either phenols or have the phenol group in their structure, which is frequently observed in lignins analyzed by this analytical technique [14].

3.2.8 Chromatographic Analysis by SEC-ELSD

The sample is solubilized in tetrahydrofuran (THF) at a final concentration of 4 μg L^{-1}. Prominence Shimadzu high performance liquid chromatography (HPLC) system can be used. Mobile phase: HPLC grade THF, flow rate 1 mL min^{-1}, analysis time of 15 min. Stationary phase: Shimpack GPC-803 column (Shimadzu), 300 mm × 8 mm; injection volume of 5 μL; column temperature 40 °C. Evaporative light scattering detector—ELSD-LT II Shimadzu; detector temperature: 40 °C; calibration standard used: Shodex XL-105 polystyrene. The analytical method will generate a result as depicted in Fig. 3.14. And the block diagram of the equipment is presented in Fig. 3.15.

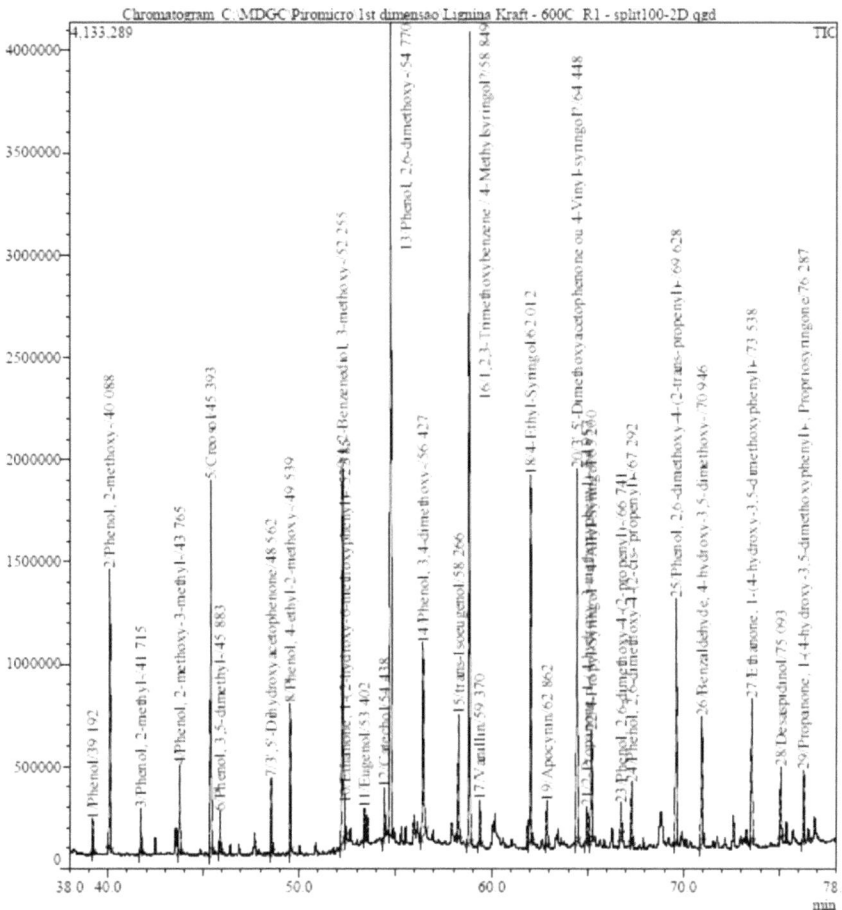

Fig. 3.12 Fragmentogram obtained by Py-GC-MS for eucalyptus Kraft lignin. Adapted from Vaz Júnior et al. (2020)

The Shimadzu's LabSolutions software can calculate the values of the analysis parameters:

(i) Mn (number average molecular mass): 569 g mol^{-1};
(ii) Mw (molecular mass): 4751 g mol^{-1}.

3.3 New Methods and Assessments

New analytical methods, based on nuclear magnetic resonance spectroscopy and mass spectrometry, which can promote an advance in the lignin structural resolution, embrace:

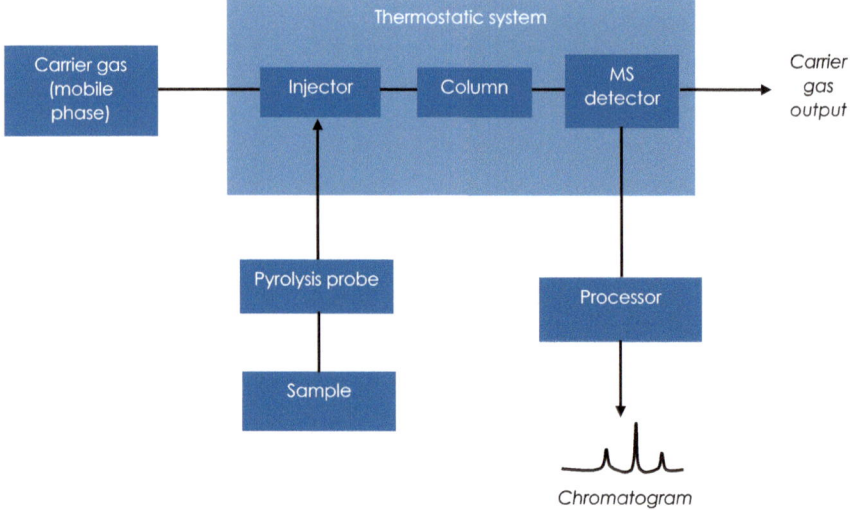

Fig. 3.13 Simplified block diagram of a Py-GC-M equipment

- The use of ^{31}P-NMR that allows the rapid and reliable identification of unsubstituted, *O*-mono substituted, and *O*-disubstituted phenols, aliphatic OH-, and carboxylic acid moieties in lignins [15].
- The use of ultra-high-resolution mass analyzers such as FT-ICR (Fourier transform ion cyclotron resonance mass analyzer) to enable lignin analysis without major sample preparation and chromatography steps [16].
- The use of two-dimensional heteronuclear single quantum coherence nuclear magnetic resonance (2D-HSQC-NMR) as a powerful tool for the elucidation of lignin interunit linkages and functionalities [17].
- And the use of matrix-assisted laser desorption/ionization-time-of-flight-mass spectrometry (MALDI-TOF MS) to complement the 2D-HSQC-NMR, in order to gain insights into linkage sequences and structural populations [17].

From these technical features, we can observe a start-of-the-art analytical chemistry dedicated to elucidate the lignin structure, their properties, and characteristics.

3.4   Green Analytical Chemistry

Instead of very useful in chemical and biochemical process to reduce their negative environmental impacts, the use of the green chemistry principle [18] addressed to analytical chemistry is a powerful tool in order to achieve more sustainable analytical processes.

The 12 principles embrace:

Table 3.3 Attribution of fragmentation peaks obtained by Py-GC-MS of eucalyptus Kraft lignin, based on similarity comparisons with the NIST spectra library. Adapted from Vaz Júnior et al. (2020)

Identification	Compound	Monomer[a]	Retention time (min)	Area	% area
2	Phenol, 2-methoxy-(guaiacyl)	G	40.088	1,677,234	5.0
4	Phenol, 2-methoxy-3-methyl	G	43.766	460,928	1.4
5	Creosol	G	45.394	1,927,232	5.7
8	Phenol, 4-ethyl-2-methoxy	G	49.538	1,077,126	3.2
9	1,2-Benzenediol, 3-methoxy	G	52.241	2,065,280	6.1
10	Ethanone, 1-(2-hydroxy-6-methoxyphenyl)	G	52.373	221,656	0.7
11	Eugenol	G	53.402	121,310	0.4
15	trans-Isoeugenol	G	58.269	616,811	1.8
17	Vanillin	G	59.371	293,997	0.9
19	Apocynin; 1-(4-hydroxy-3-methoxyphenyl)ethanone	G	62.860	338,318	1.0
21	2-Propanone, 1-(4-hydroxy-3-methoxyphenyl)-	G	64.960	402,359	1.2
1	Phenol	H	39.195	322,482	1.0
3	Phenol, 2-methyl	H	41.714	236,350	0.7
6	Phenol, 3,5-dimethyl	H	45.885	196,540	0.6
7	3',5'-Dihydroxyacetophenone	H	48.559	363,955	1.1
12	Catechol; 1,2-benzenediol	H	54.439	590,433	1.7
13	Phenol, 2,6-dimethoxy	S	54.770	5,579,131	16.5
14	Phenol, 3,4-dimethoxy	S	56.430	1,242,054	3.7
16	1,2,3-trimethoxybenzene/4-methylsyringol	S	58.848	5,222,121	15.4
18	4-Ethyl-syringol	S	62.011	2,625,986	7.8
20	3',5'-dimethoxyacetophenone or 4-vinyl-syringol	S	64.447	2,294,835	6.8
22	4-propyl-syringol + 4-allyl-syringol	S	65.216	638,049	1.9

(continued)

Table 3.3 (continued)

Identification	Compound	Monomer[a]	Retention time (min)	Area	% area
23	Phenol, 2,6-dimethoxy-4-(2-propenyl)-	S	66.733	214,819	0.6
24	Phenol, 2,6-dimethoxy-4-(2-cis-propenyl)-	S	67.292	282,667	0.8
25	Phenol, 2,6-dimethoxy-4-(2-trans-propenyl)	S	69.628	1,213,221	3.6
26	Benzaldehyde, 4-hydroxy-3,5-dimethoxy	S	70.946	856,794	2.5
27	Ethanone, 1-(4-hydroxy-3,5-dimethoxyphenyl)	S	73.539	1,165,461	3.4
28	4-(4-hydroxy-3,5-dimethoxyphenyl)propan-2-one	S	75.092	864,277	2.6
29	4-(4-hydroxy-3,5-dimethoxyphenyl)propan-1-one	S	76.286	698,215	2.1
			Total area	33,809,641	100.0

[a]*H* p-hidroxifenil; *G* guaiacyl, *S* syringil

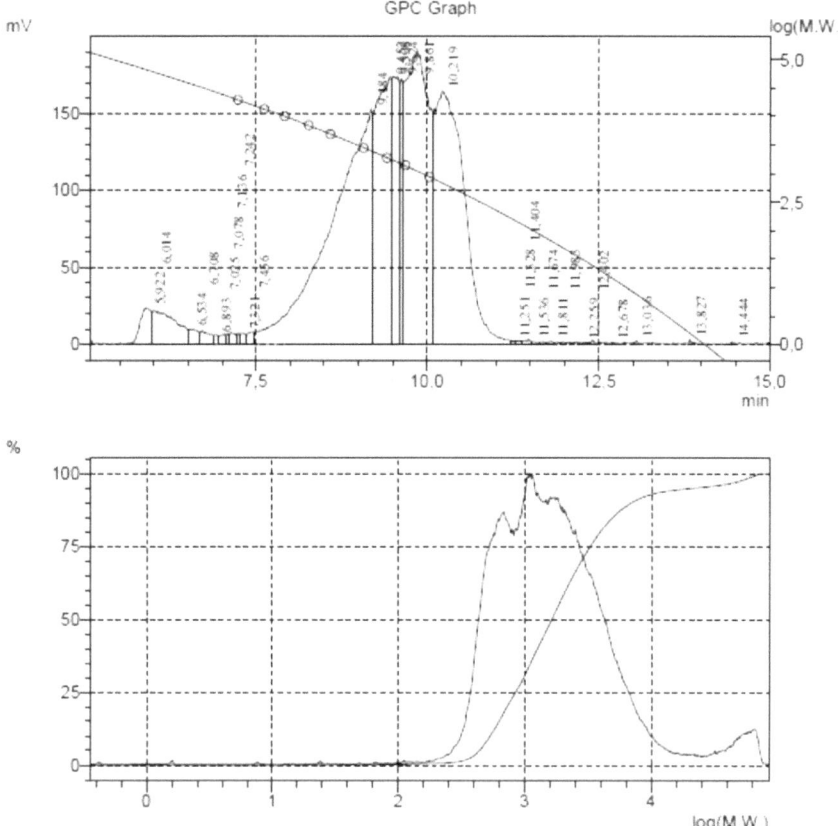

Fig. 3.14 Size exclusion chromatogram for eucalyptus Kraft lignin. The polystyrene calibration curve can be seen with 9 points in the top image. Adapted from Vaz Júnior et al. (2020)

1. **Prevention**—It is better to prevent waste than to treat or clean up waste after it has been created.
2. **Atom economy**—Synthetic methods should be designed to maximize incorporation of all materials used in the process into the final product.
3. **Less hazardous chemical syntheses**—Wherever practicable, synthetic methods should be designed to use and generate substances that possess little or no toxicity to human health and the environment.
4. **Designing safer chemicals**—Chemical products should be designed to preserve efficacy of function while reducing toxicity.
5. **Safer solvents and auxiliaries**—The use of auxiliary substances (e.g., solvents, separation agents) should be made unnecessary wherever possible and innocuous when used.

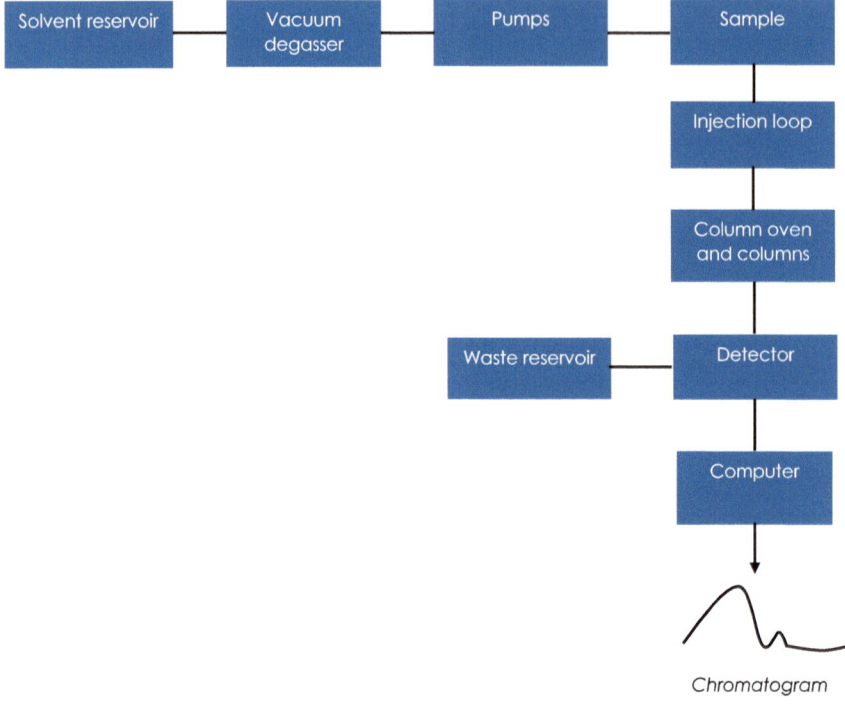

Fig. 3.15 Simplified block diagram for a SEC-ELSD equipment

6. **Design for energy efficiency**—Energy requirements should be recognized for their environmental and economic impacts and should be minimized. Synthetic methods should be conducted at ambient temperature and pressure.
7. **Use of renewable feedstocks**—A raw material or feedstock should be renewable rather than depleting whenever technically and economically practicable.
8. **Reduce derivatives**—Unnecessary derivatization (use of blocking groups, protection/deprotection, temporary modification of physical/chemical processes) should be minimized or avoided if possible, because such steps require additional reagents and can generate waste.
9. **Catalysis**—Catalytic reagents (as selective as possible) are superior to stoichiometric reagents.
10. **Design for degradation**—Chemical products should be designed so that at the end of their function they break down into innocuous degradation products and do not persist in the environment.
11. **Real-time analysis for pollution prevention**—Analytical methodologies need to be further developed to allow for real-time, in-process monitoring and control prior to the formation of hazardous substances.

12. **Inherently safer chemistry for accident prevention**—Substances and the form of a substance used in a chemical process should be chosen to minimize the potential for chemical accidents, including releases, explosions, and fires.

For analytical chemistry, we can explore the follow:

- Principle 1, preventing waste generation during the analysis.
- Principle 5, safer solvents and auxiliaries used in the analysis (e.g., for a derivatization step in gas chromatography).
- Principle 6, design analytical processes for energy efficiency.
- Principle 8, reducing analysis derivatives.
- Principle 11, using real-time analysis for pollution prevention by means the application of process analytical chemistry for lignin conversion steps.
- Principle 12, safer chemistry for accident prevention during an analytical process.

Certainly, it will guarantee more environmental-friendly analytical methods and their applications.

3.5   Conclusions

The use of various analytical techniques helps to better understand the physico-chemical properties and chemical composition of the lignin molecule, as they are complementary to each other.

The set of analytical techniques presented is suitable for evaluating the use of Kraft lignin as raw material for industrial purposes and can be easily replicated to several types of lignin for the same purpose—be it from hardwoods to softwoods—in alliance with green analytical chemistry.

Moreover, new analytical methods and assessments can contribute for more accurate analytical information regarding such a heterogeneous analytical matrix as lignin and their derived products.

References

1. M. Neureiter, H. Danner, C. Thomasser, B. Saidi, R. Braun, Dillute-acid hydrolysis of sugarcane bagasse at varying conditions. Appl. Biochem. Biotechnol. **98**, 49–58 (2002). https://doi.org/10.1385/ABAB:98-100:1-9:49
2. R.S. Fukushima, G. Garippo, A.M.Q.B. Habitante, R.S. Lacerda, Extração da lignina e emprego da mesma em curvas de calibração para a mensuração da lignina em produtos vegetais [*Extraction of lignin and use of it in calibration curves for measuring lignin in plant products*]. Revista Brasileira de Zootecnia **29**, 1302–1311 (2000). https://doi.org/10.1590/S1516-35982000000500007
3. R. Hatfield, W. Vermerris, Lignin formation in plants—the dilemma of linkage specificity. Plant Physiol. **126**, 1351–1357 (2001). https://doi.org/10.1104/pp.126.4.1351

4. M.B. Hocking, *Chemical Technology and Pollution Control*, 3ª. (Elsevier, Amsterdam, 2005). https://doi.org/10.1016/B978-0-12-088796-5.X5000-5

5. B. El Khaldi-Hansen, M. Schulze, B. Kamm, in *Qualitative and Quantitative Analysis of Lignins from Different Sources and Isolation Methods for an Application as a Biobased Chemical Resource and Polymeric Material*, ed. by S. Vaz Jr. Analytical Techniques and Methods for Biomass (Springer Nature, Cham, 2016), pp. 15–44. https://doi.org/10.1007/978-3-319-41414-0_2

6. B.C. Gambarato, Isolamento e caracterização de ligninas de palha de cana-de-açúcar [*Isolation and characterization of lignin from sugarcane straw*]. Doctoral thesis. Escola de Engenharia de Lorena, Universidade de São Paulo, Lorena, Brazil (2014). https://doi.org/10.11606/T.97.2017.tde-20112017-153808. Accessed April 2024

7. S. Brunauer, P.H. Emmett, E. Teller, Adsorption of gases in multimolecular layers. J. Am. Chem. Soc. **60**, 309–319 (1938). https://doi.org/10.1021/ja01269a023

8. N. Pareek, T. Gillgren, L.J. Jönsson, Adsorption of proteins involved in hydrolysis of lignocellulose on lignins and hemicelluloses. Biores. Technol. **148**, 70–77 (2013). https://doi.org/10.1016/j.biortech.2013.08.121

9. E.H. Novotny, M.H.B. Hayes, E.R. de Azevedo, T.J. Bonagamba, Characterisation of black carbon-rich samples by ^{13}C solid-state nuclear magnetic resonance. Naturwissenschaften **93**, 447–450 (2006). https://doi.org/10.1007/s00114-006-0126-x

10. T.A. Amit, R. Roy, D.E. Raynie, Thermal and structural characterization of two commercially available technical lignins for potential depolymerization via hydrothermal liquefaction. Curr. Res. Green Sustain. Chem. **4**, 100106 (2021). https://doi.org/10.1016/j.crgsc.2021.100106

11. J. Behin, N. Sadeghi, Utilization of waste lignin to prepare controlled-slow release urea. Int. J. Recycl. Org. Waste Agric. **5**, 289–299 (2016). https://doi.org/10.1007/s40093-016-0139-1

12. R.A. Ribeiro, S.V. Júnior, H. Jameel, H.-M. Chang, R. Narron, X. Jiang, J.L. Colodette, Chemical study of Kraft lignin during alkaline delignification of *E. urophylla* x *E. grandis* hybrid in low and high residual effective alkali. ACS Sustain. Chem. Eng. **7**, 10274–10282 (2019). https://doi.org/10.1021/acssuschemeng.8b06635

13. B.-C. Zhao, B.-Y. Chen, S. Yang, T.-Q. Yuan, A. Charlton, R.-C. Sun, Structural variation of lignin and lignin–carbohydrate complex in *Eucalyptus grandis* × *E. urophylla* during its growth process. ACS Sustain. Chem. Eng. **5**, 1113–1122 (2017). https://doi.org/10.1021/acssuschemeng.6b02396

14. J. Rencoret, J.C. Del Río, K.G.J. Nierop, A. Gutiérrez, J. Ralph, Rapid Py-GC/MS assessment of the structural alterations of lignins in genetically modified plants. J. Anal. Appl. Pyrol. **121**, 155–164 (2016). https://doi.org/10.1016/j.jaap.2016.07.016

15. D.S. Argyropoulos, N. Pajer, C. Crestini, Quantitative ^{31}P NMR analysis of lignins and tannins. J. Vis. Exp. **174**, e62696 (2021). https://doi.org/10.3791/62696

16. D.R. Letourneau, D.A. Volmer, Mass spectrometry-based methods for the advanced characterization and structural analysis of lignin: a review. Mass Spectrom. Rev. **42**, 144–188 (2021). https://doi.org/10.1002/mas.21716

17. M. Karlsson, J. Romson, T. Elder, A. Emmer, M. Lawoko, Lignin structure and reactivity in the organosolv process studied by NMR spectroscopy, mass spectrometry, and density functional theory. Biomacromol **24**, 2314–2326 (2023). https://doi.org/10.1021/acs.biomac.3c00186

18. ACS Green Chemistry Institute, 12 principles of green chemistry (2024), https://www.acs.org/greenchemistry/principles/12-principles-of-green-chemistry.html. Accessed April 2024

19. S. Vaz Júnior, W.L.E Magalhães, L.A. Colnago, W.G. de Oliveira Metodologia de caracterização físico-química de lignina kraft [Methodology for the physical-chemical characterization of kraft lignin]. Research and Development Bulletin, Brazilian Agriculture Research Corporation (Embrapa), Brasília, Brazil (2020). https://www.infoteca.cnptia.embrapa.br/infoteca/bitstream/doc/1123698/1/Boletim-de-Pesquisa-e-Desenvolvimento-Metodologia-de-Caracterizac807a771o-Fi769sico-Qui769mica-de-Lignina-Kraft-2020.pdf

Chapter 4
Extraction Processes

Abstract For laboratory study and industrial exploration of lignin as a carbona-ceous raw material or an end-product, it is desirable a high-purity with absence of impurities, which is not so easy to reach, mainly in the industry. Furthermore, it is of great interest to preserve the original chemical structure and conformation, which is even more difficult. This chapter discuss the main extraction processes to obtain lignin-types native, Kraft, Organosolv, soda and Klason, and recent advances as LignoBoost and LignoForce. Additionally, Kraft and Organosolv processes are evaluated according to green chemistry principles. And the LignoBoost process is presented as a case study.

Keywords Kraft process · Organosolv process · LignoBoost process · Separation · Purification

4.1 Description of the Most Common Processes Used in the Industry and in the R&D&I Laboratory

Extraction processes promote the lignin obtention for both analytical and industrial purposes. It is desirable a high-purity with absence of impurities, which is not so easy to reach, mainly in the industry that works with huge quantities of materials. Furthermore, it is of great interest to preserve the original chemical structure and conformation, mainly for scientific and quality purposes, which is even more difficult (as seen in Chap. 3).

The obtention of lignin in industrial or in laboratory facilities involve, in the most of cases, reactants and solvents as we can see for:

- **Native Lignin (NL) or Brauns' Lignin (BL)**: In this original procedure, plant wood is finely ground and undergoes sequential extractions with petroleum ether (an aliphatic hydrocarbon mixture), cold water, and 95% v/v of ethanol. This technique makes it possible to obtain lignin in its native form, and it is the low molecular weight part of native lignin in the plant cell wall. It is worth to remark that NL/BN shows 50% more phenolic hydroxyl groups than ball-milled wood lignin (MWL), another physical-based extraction, although the two lignins have a similar elemental composition, as observed by Cheng et al. [1].

- **Kraft Lignin (KL)**: Obtained by treating wood with sodium hydroxide and sodium hydrosulfide and, after that, acid recuperation; this lignin is widely produced by the paper and cellulose industry. Currently, Kraft pulping is globally the dominant pulping technology and accounts for 90% of chemical pulping. This is both because the product is of higher quality for most applications and the production costs are lower [2].

- **Organosolv Lignin (OL)**: Lignin can also be extracted from lignocellulosic fibers, such as sugarcane bagasse, coconut fiber, and oil palm mesocarp fiber. This approach uses organic solvents (e.g., ethanol) to generate purer, sulfur-free lignins, making it a sustainable alternative, for example, for the synthesis of phenolic compounds. Up to 77.9% of the lignin present in the biomass can be recovered via the Organosolv process, with high-purity, low molecular weight, and higher total phenolic OH and lower aliphatic OH contents [3].

- **Soda Lignin (SL)**: Soda pulping process involves sodium hydroxide for cooking chemical, with small quantities of phenolic hydroxyls formed and primary aliphatic hydroxyl decreasing through the cleavage of aryl-ether linkages. SL is considered as sulfur-free lignin which resembles more closely the structure of NL/BL [4].

- **Klason Lignin (KSL)**: The insoluble residue portion after removing the ash by concentrated acid hydrolysis of the plant tissues, to determine the lignin content in plants by a gravimetric method for determining lignin directly in woody materials [5].

Generally, KL and SL are the main sources of technical grade lignin industrially produced. OL is under development, and NL/DL and KSL are especially useful for analytical purposes.

Additionally, there are more recent developed industrial processes addressed to lignin recovery from Kraft industrial streams:

- **WestVaco Process**: This process involves the isolation of lignin from Kraft black liquor, and aims to extract lignin efficiently from pulping process streams. This process was patented by the West Virginia Pulp and Paper Company in 1949 and is still in use to produce the indulin lignin, which is highly functionalized in carboxylic acid, aliphatic, and aromatic hydroxyl groups, and is free of reducing sugars [6].

- **LignoBoost Process**: Another process for lignin isolation also targets the Kraft black liquor. It is a patented extraction process divided into two steps: a separation step, followed by a washing step. Dividing the extraction into two steps makes it possible to optimize the conditions in each step and produce a very high-quality lignin [7].

- **LignoForce System™**: This process is other approach for isolating lignin from Kraft black liquor. It has been compared to the WestVaco process in terms of product quality. It is a patent-pending process in which the black liquor is oxidized under controlled conditions before the acidification step of conventional lignin recovery processes [8].

- **Sequential Liquid-Lignin Recovery and Purification (SLRP)**: The patented SLRP™ process precipitates lignin from black liquor as a true liquid phase that separates by gravity. This is different from traditional processes that precipitate lignin as small solid particles that have to be filtered. CO_2 and H_2S generated are recycled and reacted with the incoming black liquor, reducing the carbon dioxide consumption by 30%; supposedly, it is the lowest cost and most energy efficient process for recovering lignin from pulping residuals [9].

It is very difficult to establish the best lignin recovery process because it depends on of aspects as CAPEX,[1] OPEX,[2] raw material availability and target products. In order to help clarify it, Kienberger et al. [10] published a complete evaluation of these four-isolation processes in order to obtain technical lignin highlighting the technical and economic aspects of each one.

4.2 A Critical Evaluation of Processes: Advantages Versus Disadvantages According to Green Chemistry Principles

As previously introduced, several processes can be applied to lignocellulosic biomass to fractionate it and release its components, i.e., cellulose, hemicellulose, and lignin. This is the *modus operandi*, for instance, of the pulp and paper industry and cellulosic fuels (e.g., ethanol).

Considering lignin extraction or isolation for industrial purposes, we can take into account the characteristics of Kraft process (the main) and the Organosolv process (the promise). From these characteristics we can infer the advantages and disadvantages for each one according the green chemistry principles [11], in order to reach more eco-friendly processes.

4.2.1 Kraft Process

The Kraft process (Figs. 4.1) can be briefly described through the main sequential steps:

- **Impregnation**: Wood chips, typically 12–25 mm (0.47–0.98 inches) long and 2–10 mm (0.079–0.394 in.) thick, are used in pulp production. These chips undergo

---

[1] CapEx = capital expenditure, How to Calculate CapEx - Formula (corporatefinanceinstitute.com).

[2] OpEx = operational expenditure, Operating Expenses—Definition, Example, Type, Explain (corporatefinanceinstitute.com).

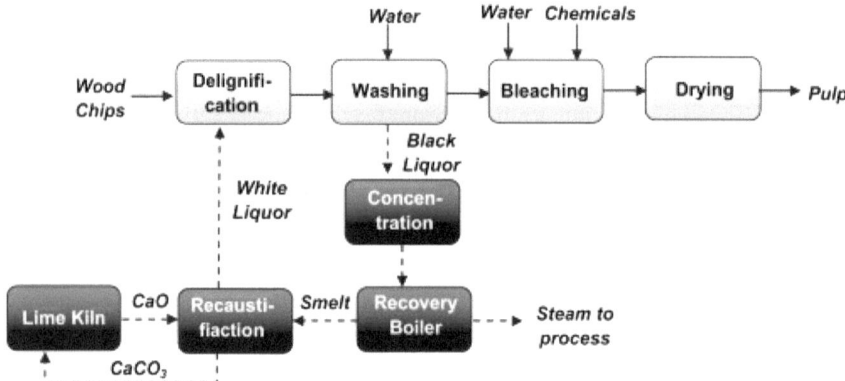

Fig. 4.1 A simplified flowchart of the Kraft process, with lignin been recovered by the acidification of the black liquor. *Source* Mateos-Espejel et al. [13]. Reprinted with permission from Elsevier

presteaming, where they are wetted and preheated with steam. The steam treatment causes air within the chips to expand, resulting in the expulsion of about 25% of the air. The chips are then saturated with a mixture of *black liquor* (a combination of sodium hydroxide, NaOH, and sodium sulfide, Na_2S, containing dissolved lignin) and *white liquor* (a strongly alkaline solution mainly of sodium hydroxide, and minor quantities of salts as sodium carbonate, sodium sulfate, sodium thiosulfate, and other non-processed compounds). Impregnation ensures that the cooking liquor penetrates the capillary structure of the chips, allowing low-temperature chemical reactions with the wood to begin. Approximately 40–60%wt/wt. of the total alkali consumption occurs during impregnation in the continuous process.

- **Cooking**: The impregnated wood chips are cooked in pressurized vessels called digesters. The cooking liquor consist of a blend of white liquor, water within the chips, condensed steam, and weak black liquor. During cooking, the bonds linking lignin, hemicellulose, and cellulose breakdown, separating the cellulose fibers. The result is wood pulp, which forms the basis for paper production.

- **Lignin Recuperation**: Basically, it is reached by means the use of considerable amount of inorganic acids like sulfuric acid added to the black liquor followed by filtration; however, an alternative is the use of non-toxic aluminum potassium sulfate dodecahydrate, $AlK(SO_4)_2 \cdot 12H_2O$ [12].

Regarding the black liquor, the source from that the lignin (Fig. 4.2) is recovered, it is worth to consider a production of 1 to 1.4 kg of concentrate liquor per 1 kg of cellulose with almost of 7%wt/wt. of lignin. After its concentration, it presents high viscosity and high calorific power—due to lignin presence—near by 13,400 kJ kg^{-1} for its burning into the boiler to generate electricity for the energy self-sufficiency industrial plant.

Some chemical reactions related to the Kraft processing can be seen in the Scheme 4.1.

Fig. 4.2 A eucalyptus (hydrides *Eucalyptus grandis* × *Eucalyptus urophylla*) lignin obtained from Kraft process

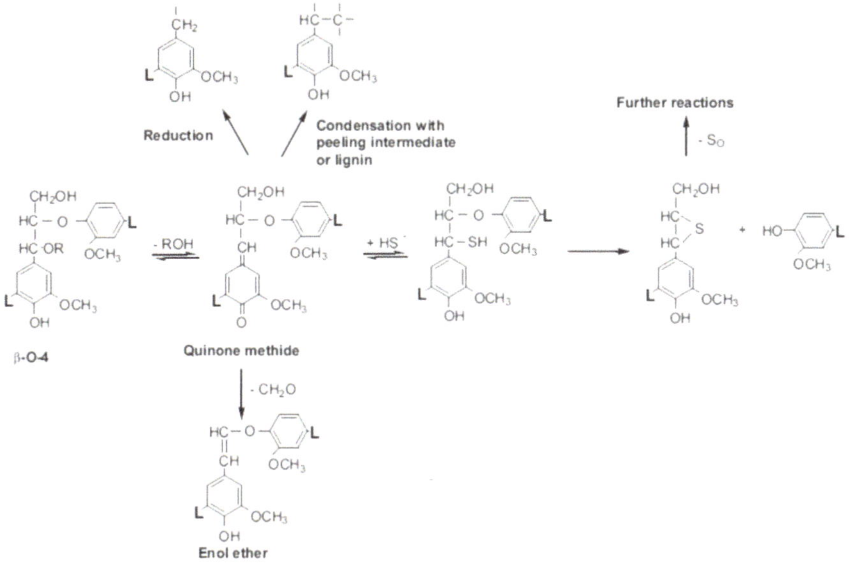

Scheme 1 Chemical reactions of phenolic β-O-4 linkages in the lignin structure occurring during the Kraft processing for wood pulping

The advantage of Kraft process according to the green chemistry principles [11] are related to the production of a renewable feedstock (i.e., Kraft lignin) (principle 7) to be used by the chemical industry and design for energy efficiency (as cited before, lignin has a remarkable calorific value) (principle 6). As disadvantages, they are related to the production of H_2S and SO_2 gases, which are flammable, acute toxic and environmental hazard, and corrosive and acute toxic, respectively [14], against the principle 12 related to inherently benign chemistry for accident prevention.

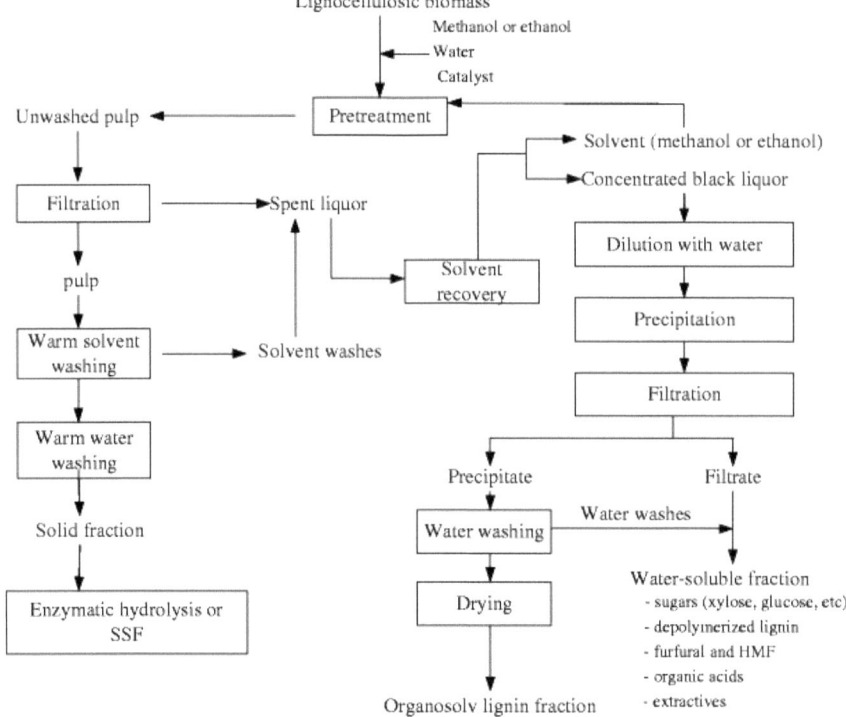

Fig. 4.3 A simplified flowchart of Organosolv process for lignocellulosic biomass. *Source* Zhao et al. [17]. Reprinted with permission from Springer Nature

4.2.2 Organosolv Process

Regarding to the Organosolv process (Fig. 4.3), it can be briefly described through the main characteristics:

- Organosolv involves treating a lignocellulosic feedstock (like chipped wood or sugarcane bagasse) with an aqueous organic solvent—ethanol is the most common—at temperatures ranging from 140 °C to 220 °C.
- This process breaks down lignin by hydrolytic cleavage of α-aryl-ether links, resulting in fragments that are soluble in the solvent system (Fig. 4.4).
- Common solvents used include acetone, methanol, ethanol, butanol, ethylene glycol, formic acid, and acetic acid.
- Solvent concentrations in water typically range from 40 to 80% v/v.
- Ethanol is often preferred due to its cost-effectiveness and ease of recovery, although other solvents like acetone and butanol have also been explored.
- Generally, an inorganic acid (e.g., H_2SO_4) is added to act as catalyst.
- Organosolv allows the recovery of relatively high-quality lignin (Fig. 4.5), adding value to a process stream that would otherwise be considered waste.

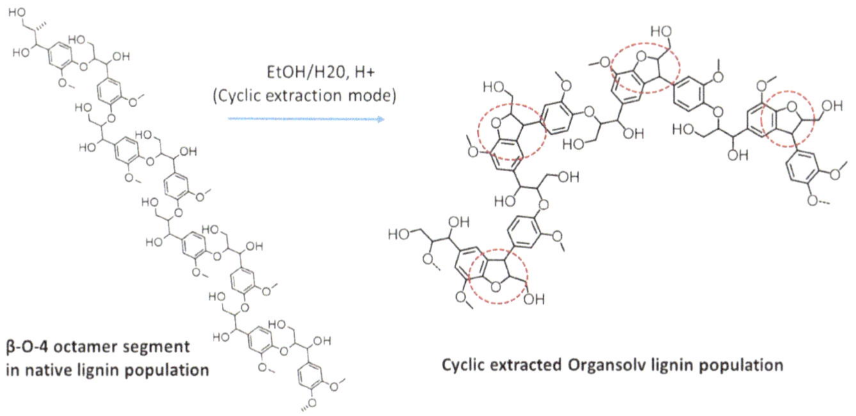

Fig. 4.4 Reactivity of a homogeneous octameric segment consisting of only β-O-4′ linkages in native lignin. An intramolecular condensation is identified that occurs alternately in the β-O-4′ sub-units. The alternate occurrence is marked with red dotted circles. *Source* Karlsson et al. [18]. Reprinted with permission from the American Chemical Society

Fig. 4.5 A sweet grass (*Hierochloë hirta* subspecies *arctica*) lignin obtained by the Organosolv process

• Solvent Recovery: Solvents can be easily recovered by distillation, leading to less water pollution and eliminating the odor associated with Kraft pulping.

Certainly, Organosolv processes—e.g., Alcell process [15] and Lignol process[3]— are very promising for lignocellulosic biomasses as sugarcane and grasses (e.g., sweet grass and elephant grass) due to its facility to fractionate them, as suggested by Tofani et al. [16] for several biomasses.

As advantage of Organosolv process according to the green chemistry principles [11], we can cite the production of a renewable feedstock (principle 7) to be used by

---

[3] https://www1.eere.energy.gov/bioenergy/pdfs/ibr_demonstration_lignol.pdf.

the chemical industry, and safer solvents and auxiliaries (principle 5). On the other hand, as disadvantages: reduced lignin yield (5–7%wt/wt) when compared against Kraft process (20–30%wt/wt).

4.3   Discussion of a Case Study

The LignoBoost process is a good example of lignin recovery for a case study. As previously commented, this process recovers the lignin from the black liquor from the Kraft processing of wood for the paper and pulp industry. Figure 4.6 depicts the process.

The steps involved in this process and their unitary operations are [7]:

- **Step 1—Separation**: The first step in the LignoBoost process is to separate the lignin from the mill's black liquor. Black liquor is taken from the evaporation, and the pH value is lowered with CO_2—by the generation of carbonic acid and its deprotonation—and gas from the second step of the LignoBoost process. When the pH value drops, lignin precipitates and it is separated from the liquor with a press filter. The output of the first step is LignoBoost crude lignin.
- **Step 2—Washing**: In the second step, the lignin is purified. The crude lignin is washed in acid solution and then dewatered in a second filter press. The lignin purity is highly dependent on the conditions in the washing step. LignoBoost is engineered to produce a high-pure lignin.

Other associated hot-spots for the process are:

- Filter presses are key for the quality of the extracted lignin.
- Return flow to the mill with minor changes to the mill's operation.
- Pulp mill sulfur balance.
- High operation safety ensured.
- Safe gas handling to reduce CO_2 consumption.

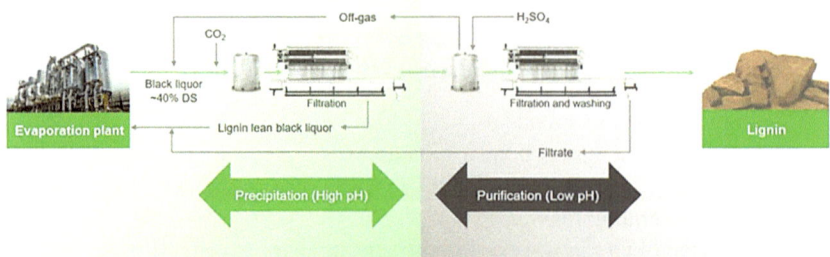

Fig. 4.6 The LignoBoost process and its steps to obtain lignin from black liquor from the Kraft wood processing. *Source* Adapted from Valmet [7]. Reprinted with permission from Valmet

- Handling lignin in the mill for stored, package and easy transportation.
- Lignin as fuel in the lime kiln.

According to Valmet [7], up to 70%wt/wt of lignin can be recovery by the LignoBoost process.

Additionally, we can note that the LignoBoost process can promote the green chemistry principle 7 [11], related to the high-purity lignin as renewable feedstock for industry.

4.4 Conclusions

The extraction and lignin recovery or isolation are paramount technological steps addressed to obtain technical lignin in order to be used as a carbonaceous raw material in industrial processing.

Several processes can be applied to lignocellulosic biomass in order to fractionate it and release its components, i.e., cellulose, hemicellulose, and lignin. However, the Kraft process is the most used by the pulp & paper industry—the main source of technical lignin—being able to highlight the Organosolv process for sugarcane, sweet grass, miscanthus, among other grasses.

After the release of lignocellulosic components, the lignin should be recovered or isolated by a further process. Generally, it is obtained from the black liquor, in the case of Kraft process, for which the LignoBoost is a good case study discussed in this chapter.

References

1. X.-C. Cheng, X.-R. Guo, Z. Qin, H.-M. Liu, J.-R. He, X.-D. Wang, Sequential aqueous acetone fractionation and characterization of Brauns native lignin separated from Chinese quince fruit. Int. J. Biol. Macromol. **201**, 67–74 (2022). https://doi.org/10.1016/j.ijbiomac.2021.12.114
2. D.D.S. Argyropoulos, C. Crestini, C. Dahlstrand, E. Furusjö, C. Gioia, K. Jedvert, G. Henriksson, C. Hulteberg, M. Lawoko, C. Pierrou, J.S.M. Samec, E. Subbotina, H. Wallmo, M. Wimby, Kraft lignin: a valuable, sustainable resource, opportunities and challenges. Chemsuschem **16**, e202300492 (2023). https://doi.org/10.1002/cssc.202300492
3. G.-H. Kim, B.-H. Um, Fractionation and characterization of lignins from *Miscanthus* via organosolv and soda pulping for biorefinery applications. Int. J. Biol. Macromol. **158**, 443–451 (2020). https://doi.org/10.1016/j.ijbiomac.2020.04.229
4. J. Sameni, S.A. Jaffer, M. Sain, Thermal and mechanical properties of soda lignin/HDPE blends. Compos. A Appl. Sci. Manuf. **115**, 104–111 (2018). https://doi.org/10.1016/j.compos itesa.2018.09.016
5. P. Bajpai, in *Chapter 19—Pulp Bleaching*, ed. by P. Bajpai. *Biermann's Handbook of Pulp and Paper* (Third Edition) (Elsevier, 2018), pp. 465–491. https://doi.org/10.1016/B978-0-12-814 240-0.00019-7
6. Ingevity, Products/Indulin® AT (2024), https://www.ingevity.com/products/indulin-at-agricu ltural-chemicals/. Accessed April 2024

7. Valmet, LignoBoost®—the process (2024), https://www.valmet.com/pulp/other-value-adding-processes/lignin-extraction/lignoboost-process/. Accessed April 2024
8. FPInnovations, LignoForce System (2024), https://web.fpinnovations.ca/lignin-from-black-liquor/. Accessed April 2024
9. Liquid Lignin, About us (2024), https://liquidlignin.com/about-us/. Accessed April 2024
10. M. Kienberger, S. Maitz, T. Pichler, P. Demmelmayer, Systematic review on isolation processes for technical lignin. Processes **9**, 804 (2021). https://doi.org/10.3390/pr9050804
11. ACS Green Chemistry Institute, 12 principles of green chemistry (2024), https://www.acs.org/greenchemistry/principles/12-principles-of-green-chemistry.html. Accessed April 2024
12. N.D. Luong, N.T.T. Binh, L.D. Duong, D.O. Kim, D.-S. Kim, S.H. Lee, B.J. Kim, Y.S. Lee, J.-D. Nam, An eco-friendly and efficient route of lignin extraction from black liquor and a lignin-based copolyester synthesis. Polym. Bull. **68**, 879–890 (2012). https://doi.org/10.1007/s00289-011-0658-x
13. E. Mateos-Espejel, L. Savulescu, F. Maréchal, J. Paris, Systems interactions analysis for the energy efficiency improvement of a Kraft process. Energy **35**, 5132–5142 (2010). https://doi.org/10.1016/j.energy.2010.08.002
14. PubChem, Explore chemistry (2024), https://pubchem.ncbi.nlm.nih.gov/. Accessed April 2024
15. W.H. Klingensmith, J.H. Lora, M.J. Trojan, High purity lignin derivatives for use in rubber. Australian patent number AU652561B2 (1991), https://patentimages.storage.googleapis.com/6d/3e/f4/4a0a191418787e/AU652561B2.pdf. Accessed April 2024
16. G. Tofani, E. Jasiukaitytè-Grojzdek, M. Grilc, B. Likozar, Organosolv biorefinery: resource-based process optimization, pilot technology scale-up and economics. Green Chem. **26**, 186–201 (2024). https://doi.org/10.1039/D3GC03274D
17. X. Zhao, K. Cheng, D. Liu, Organosolv pretreatment of lignocellulosic biomass for enzymatic hydrolysis. Appl. Microbiol. Biotechnol. **82**, 815–827 (2009). https://doi.org/10.1007/s00253-009-1883-1
18. M. Karlsson, J. Romson, T. Elder, A. Emmer, M. Lawoko, Lignin structure and reactivity in the organosolv process studied by NMR spectroscopy, mass spectrometry, and density functional theory. Biomacromol **24**, 2314–2326 (2023). https://doi.org/10.1021/acs.biomac.3c00186

Chapter 5
Chemical Processing and Their Products

Abstract Conversion processes generally related to chemocatalysis, molecular deconstruction or depolymerization, thermochemistry, and structural modifications can boost the development of high-value products from lignin. In this chapter, these classes of chemical conversion processes will be treated in order to explore all possibilities to achieve several organic chemicals. A case study base on lignin cracking is also presented. And the evaluation of advantages and disadvantages according the green chemistry principles is considered.

Keywords Chemocatalysis · Cracking · Thermochemistry · Structural modifications · Technology readiness level · Biorefineries

5.1 Description of the Most Common Processes Used in the Industry and Those in Development Based on Chemical Transformation

The use of lignins and their derivatives to obtain chemicals, fuels, and materials involve the application of technologies and conversion processes generally related to chemocatalysis, depolymerization, thermochemistry, and structural modifications. These approaches can boost the development of high-value products in order to promote green production chains.

The *technology readiness level* (TRL) for lignin conversion techniques varies, but ongoing research aims to improve efficiency and scalability. For instance, the TRL lignin-based phenolic resins—obtained by means the lignin catalytic cracking—are very developed and are actually in TRL-8 [1].

Considering an industrial innovation, the successful lignin valorization can contribute to sustainable development and address energy and environmental challenges. Furthermore, these advancements are critical for transitioning toward a more sustainable and bio-based economy (or bioeconomy).

S. Vaz Jr., *The Lignin Macromolecule*,
SpringerBriefs in Applied Sciences and Technology,
https://doi.org/10.1007/978-3-031-75511-8_5

5.1.1 Chemocatalysis

Chemocatalysis plays a crucial role in the valorization of lignin in current lignocellulosic *biorefineries*. As lignin presents challenges due to its recalcitrant nature, the use
of catalysts is paramount to produce mainly chemicals from this raw carbonaceous
material. Implementing renewable compounds will enhance the sustainability of the
chemical industry and open doors to innovative products based on unique biobased
structures.

By facilitating chemical reactions, catalysts enable the transformation of lignin's
complex structure into simpler, more useful molecules. In summary, chemocatalytic
strategies offer exciting opportunities for transforming lignin into valuable chemicals,
contributing to a more sustainable future in the chemical industry.

When we consider the main catalyst classes, i.e., *heterogenous* and *homogeneous*,
it is highly desirable that catalysts should be renewable, recyclables, and recoverable
in order to guarantee the economic viability of chemical conversion processes. Thus,
from these point of view heterogeneous catalysts may be more applicable. However,
Lin et al. [2] highlighted the need to take into account two aspects for the application:

(i)   The precise role of active sites.
(ii) The catalyst–substrate chemistry that underpins the catalytic activity.

Figure 5.1 depicts heterogeneous catalysts for chemical conversion of lignocellulosic biomass, considering some types of chemical reactions.

Lewis acids and Brønsted acids are very important for chemocatalysis. A Lewis
acid is an electron-pair acceptor and therefore able to react with a Lewis base to
form a Lewis adducts. And a Brønsted acid is a proton donate to a base or to the

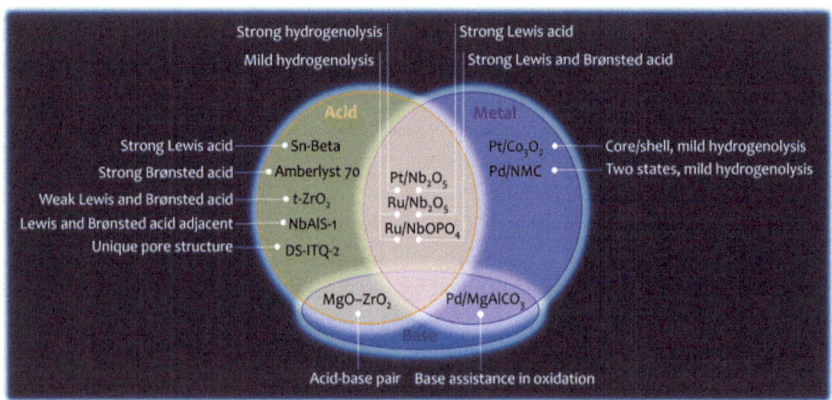

Fig. 5.1 An overview of the activity of selected heterogeneous catalysts for chemical conversion
of lignocellulosic biomass into chemicals. *Source* Lin et al. [2]. Reprinted with permission of the
Royal Society of Chemistry

corresponding chemical species. These mechanisms shape the kinetic effect on the reaction to be catalyzed.

In a practical way, chemocatalysis is directly applied to molecular deconstruction and to thermochemistry, to be seen in the next items.

5.1.2 Molecular Deconstruction

The lignin molecular deconstruction sometimes is known as *depolymerization* or *lignin cracking*. It is also a mode of application of the chemocatalysis with or without association to biological processes (i.e., enzymes or microbes) (to be seen in Chap. 6) in order to obtain products derived from the lignin macromolecular structure.

However, there is a real challenge in lignin decomposition to be overcome: the lignin fragment repolymerization, or the recombination in order to restore the initial structure. The conversion of lignin into products cannot reach good yields by the reason of the occurrence of two concurrent processes, i.e., *depolymerization* (the molecular deconstruction) and *repolymerization* of lignin products (such as condensation of phenolic moiety and the reactive groups like carbonyls) [3]. The controlling of the repolymerization can be reached by means:

- Redox reactions, as the irreversible lignin oxidation, e.g., the use of Fenton reactant (Fe(III) and peroxide) to obtain monomers; and the lignin reduction with metal catalysts (e.g., Ni, Co, Ru, Mo) under high temperature and pressure or residence time.
- Solvents (e.g., alcohols), co-solvents (e.g., water/phenol), and Lewis acids ($CrCl_3$, $FeCl_3$, $NiCl_2$, $ZnCl_2$) for lignin cracking.
- Capping agents, as boric acid, phenols, and formic acid, have shown evidence to suppress repolymerization.
- Oxides of metal and non-metal like zeolites, SiO_2, and Al_2O_3 assist in the conversion of lignin into aromatics.

The hydrogenolysis reaction associated with photochemistry and catalysis can cleave, for instance, the β-O-4 alcohol into ketones and phenol [4]. It is depicted in Scheme 5.1.

The production of aromatic compounds from lignin derivatives can be reached by the cracking of phenol moieties (guaiacol-like) - under the presence of pristine amorphous SiO_2–Al_2O_3 as a catalyst and methanol as solvent—generating benzene-homologues by alkylation [5]. It can be seen as the way to substitute fossil raw material, i.e., oil, by carbonaceous renewable raw material, i.e., lignocellulosic raw material.

Furthermore, the production of aromatic strategic molecules from lignin depolymerization or catalytic cracking strategic molecules—using hydroprocessing, decarboxylation, dealkylation, demethoxylation reactions—can generate platform molecules and drop-in chemicals, as proposed by Sudarsanam et al. [6].

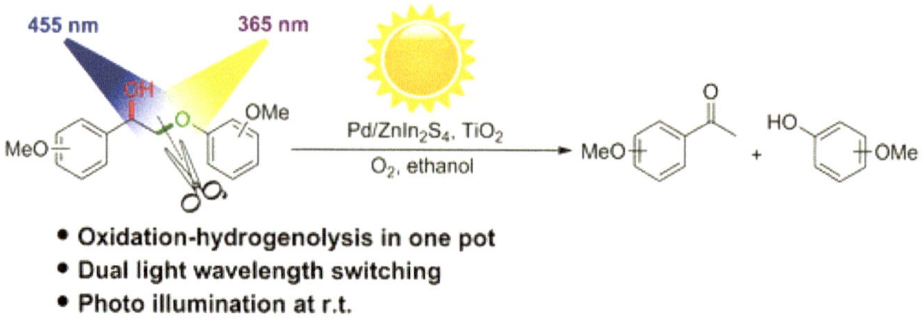

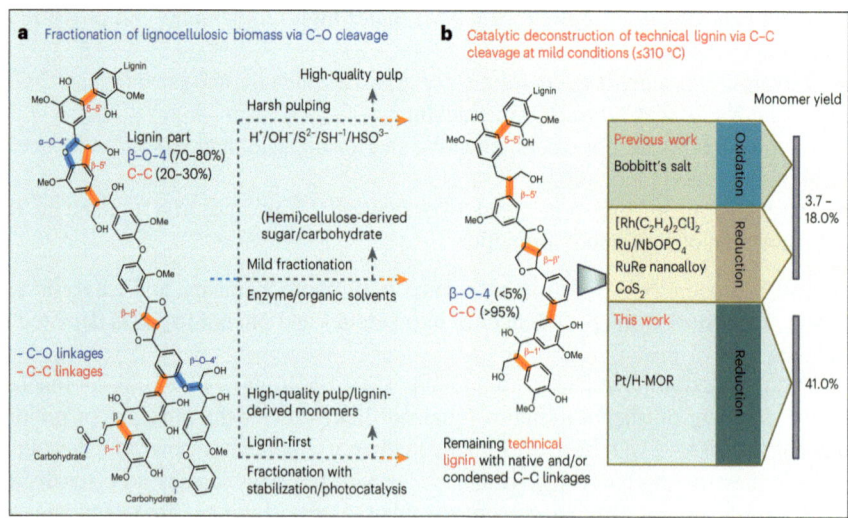

- **Oxidation-hydrogenolysis in one pot**
- **Dual light wavelength switching**
- **Photo illumination at r.t.**

Scheme 5.1 Photocatalytic cleavage of β-O-4 alcohols into ketones and phenols by oxidation–hydrogenolysis reaction via DLWS (dual light wavelength switching) strategy. *Source* Luo et al. [4]. Reprinted with permission from the American Chemical Society

The C–C bonds presented in the lignin structure can limit monomer yields from current depolymerization strategies, which mainly target C–O bonds. Some approaches can be used to overcome this restriction, as the use of a Pt (de)hydrogenation function leads to olefinic groups close to recalcitrant C–C bonds, which can undergo β-scission over zeolitic Brønsted acid sites [7]. Figure 5.2 depicts a general overview of available approaches to be applied to the lignin cracking to producer monomers.

Fig. 5.2 State-of-the-art approaches for the fractionation of lignocellulosic biomass via C–O cleavage (**a**); the C–C and C–O bonds in lignin are highlighted in red and blue, respectively. And current and proposed catalytic approaches for the deconstruction of technical lignin via C–C cleavage under mild conditions (≤ 310 °C) (**b**). *Source* Luo et al. [7]. Reprinted with permission from Springer Nature

As a state-of-the-art R&D&I example, we can cite the production of lignin-based jet fuel (LJF)—a SAF, sustainable aviation fuel—composed of C6-C18 mono-, di-, and tri-cycloalkane that meet the ASTM D7566 norm for required properties for jet fuel [8]. This LJF was produced by means the lignin degradation in monomers and dimers, using a zeolite and noble metal-catalyst (Ru/Al_2O_3), with posterior coupling and hydrodeoxygenation, generating cyclic structure hydrocarbons.

5.1.3 Thermochemistry

Some thermochemical processes to be applied to lignin in order to obtain chemicals and materials are *pyrolysis* and *hydrothermal carbonization*. From the first one we can obtain, for instance, biochar and bio-oil, and from the second one hydrochar and phenol derivatives.

- **Pyrolysis**: It is a thermal decomposition process that typically takes place in a fluidized bed reactor under oxygen-free conditions. Based on the heating rates, pyrolysis can be classified as *slow* and *fast*. Significant products generated during lignin pyrolysis include volatile organic compounds (VOCs) such as methanol, ethanol, acetone, acetaldehyde, and phenol derivatives like guaiacol, syringol, catechol, and other mono-lignols. Gaseous compounds such as hydrogen, methane, ethane, ethene, carbon monoxide, and carbon dioxide are also produced during lignin pyrolysis. In addition to producing biorenewable energy products such as bio-oil and synthesis gas (syngas), large-scale biochar production—the solid product—is feasible from the lignin pyrolysis process. Biochar is a thermally stable solid material, which can have a large surface area known for possessing abundant hydroxyl and carbonyl groups that can undergo further functionalization processes [9].
- **Hydrothermal carbonization**: It takes place under high temperatures and in liquified conditions to transform both the physical and chemical characteristics of biomass. In the hydrothermal carbonization of lignin, the β-O-4 linkages and C–C bonds are cleaved from 330 to 400 °C and pressure of 50–115 MPa; however, C–C bonds in the aromatic rings are not affected. At lower temperatures and shorter times, the hydrothermal carbonization of lignin produces both phenolic monomers and dimers via primary hydrolysis and cleavage of ether and aliphatic bonds. After prolonged times and at high temperatures, demethoxylation and alkylation reactions of phenolic materials occur leading to the production of several alkyl phenols [9]. An advantage of this process is to produce structured carbonaceous material to be used as activated carbon for adsorption technologies [10] and a solid fuel (hydrochar) [11].

Additionally, catalytic fast pyrolysis, using zeolites (e.g., ZSM-5-type zeolites) can be aimed at reducing the oxygen content of the bio-oils to reduce the level of deoxygenation required during hydrotreatment to produce hydrocarbon fuels [12]. Figure 5.3 show some reaction pathways for catalytic pyrolysis of lignin.

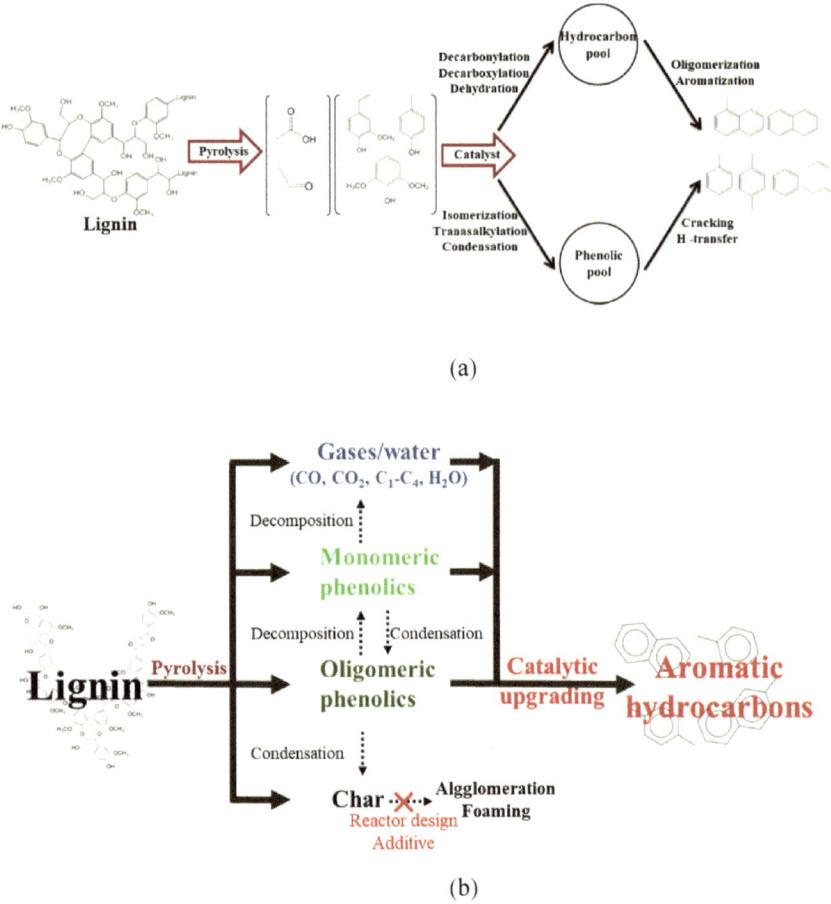

(a)

(b)

Fig. 5.3 Reaction pathways of catalytic pyrolysis of lignin (**a**) aromatic hydrocarbons formation (**b**) overall. *Source* Ha et al. [13]. Reprinted with permission from Elsevier

As all thermochemical processes, pyrolysis and hydrothermal characterization will consume a considerable energy to reach those stated products.

5.1.4 Structural Modifications

Some classes of reactions can be used in organic synthesis routes aiming at the structural modification of the lignin surface, in order to improve the surface properties that will provide the controlled release of biologically active molecules (e.g., pesticides). In other words, inserting chemical groups, such as methoxyls, carbonyls and unsaturated alkyls, for the formation of intermolecular lignin-active ingredient interactions.

As an example, Vaz Jr. and Salvador [14] developed an innovative process for the structural modification of Kraft lignin followed by the incorporation of the herbicide picloram to obtain a nanoformulation for controlled release, by means of a hydroxymethylation reaction under continuous-flow regime. It reached a TRL-4—that means, it was validated in laboratory scale. Scheme 5.2 presents the hydroxymethylation reaction of lignin modification.

Acetylation [15], methylation [16], and amination [17] are other examples of reactions used to modify the lignin structure in order to insert functional groups to improve the functionality of the macromolecule. These reactions are depicted in Schemes 5.3, 5.4, and 5.5.

Modifications presented in Schemes 5.2, 5.3, 5.4, and 5.5 can be useful to produce carriers for drug and agrochemical delivery systems.

Additionally, the production of epoxide functionalized lignin with lignin-based polyepoxides are a modification-type mostly used as cross-linkers for the introduction of hard segments into epoxy resins [18].

Structural lignin modification can occur also by means the use of thermochemical process (e.g., hydrothermal carbonization), as previously introduced.

However, the most common and simple lignin modification is to produce ligno-sulphonates. Lignosulfonates are macromolecules derived from the native structure of lignin, with characteristics of a polyelectrolyte, soluble in water and with an anionic character. Its fundamental unit is of the phenylpropane-type and has several methoxyl, hydroxyl, and carboxyl radicals (Fig. 5.4). These compounds are obtained

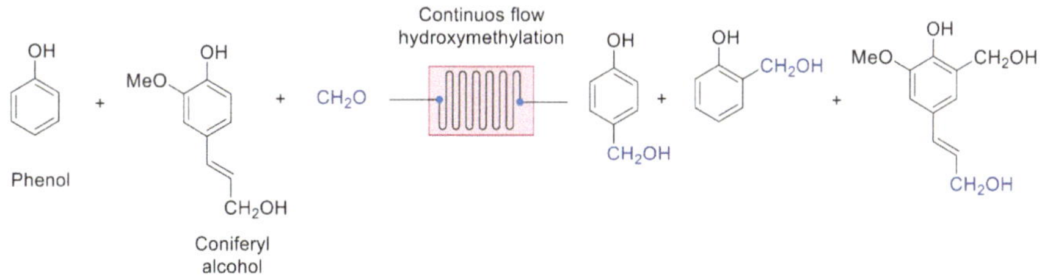

Scheme 5.2 Proposed hydroxymethylation reaction for Kraft lignin modification. *Source* Vaz Jr. and Salvador [14]. Reprinted with permission from Elsevier

Scheme 5.3 Acetylation of the lignin macromolecule

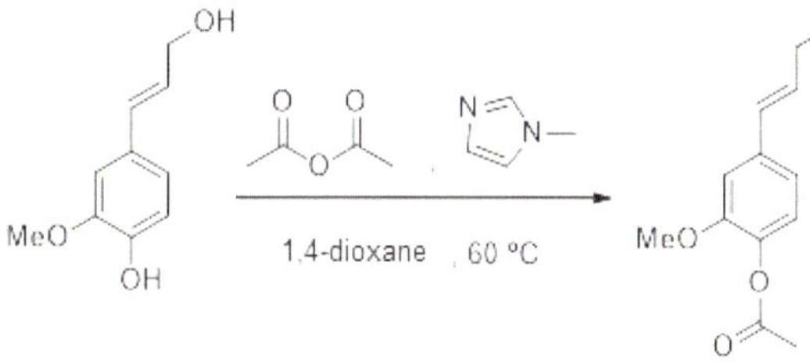

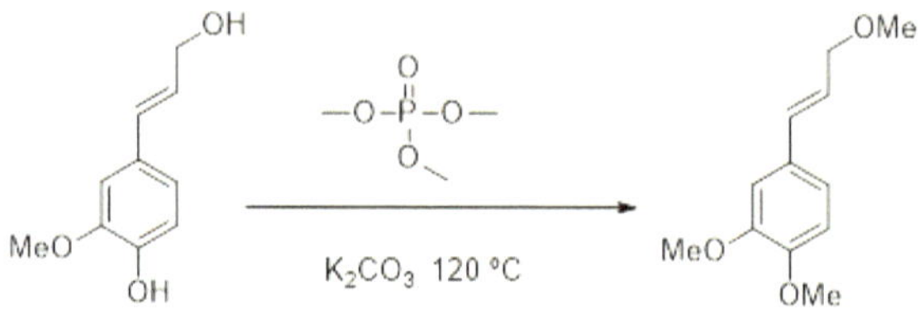
OH
O
—O-P-O—
O
K2CO3 120 °C
MeO
OH
OMe
MeO
OMe

Scheme 5.4 Methylation of the lignin macromolecule

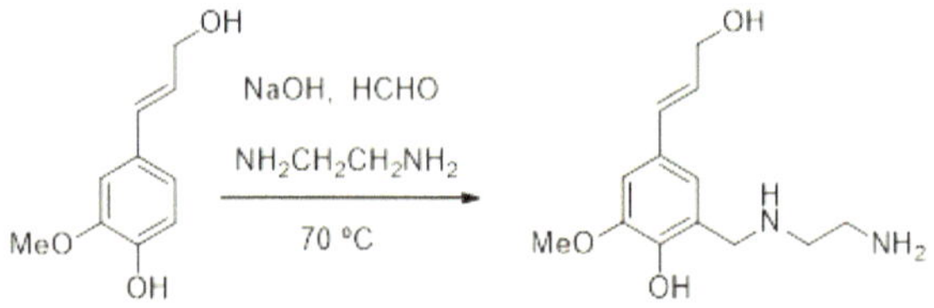
OH
NaOH, HCHO
NH2CH2CH2NH2
70 °C
MeO
OH
OH
H
N
NH2
MeO
OH

Scheme 5.5 Amination of the lignin macromolecule, by means the Mannich reaction

from the process of separating cellulose and lignin from wood (sulfite processing of wood).

Lignosulfonates are used in construction industry, water treatment plants, dispersants, textile industry, and chemical industry, as presented in Chap. 1 (item 1.4).

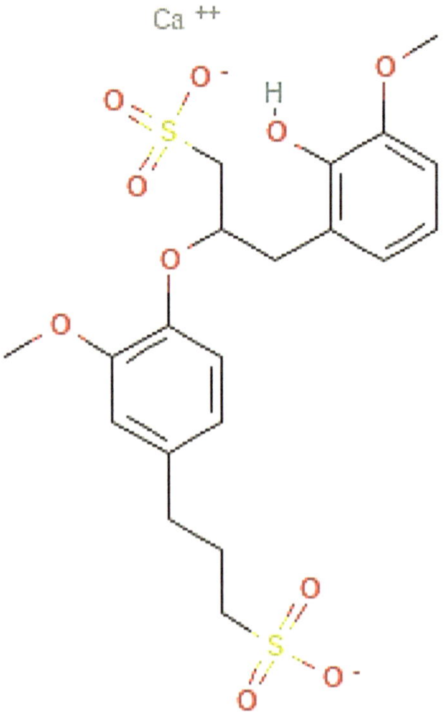
Ca ++

Fig. 5.4 Chemical structure of calcium lignosulfonate

5.2 Discussion of a Case Study

Based on the known reactivity of lignin, cracking tests were carried out in a Parr high-pressure stainless-steel reactor. 2 g of lignin in 20 mL of water was used as starting material, at 200 °C for 2 h. These tests were repeated changing the standard reaction conditions, according to Table 5.1. The lignins used in the reactions were a standard sulphonated lignin (Sigma-Aldrich) and the lignins extracted from fines and Kraft black liquor from pine and eucalyptus. The cracking products were analyzed by ultra-performance liquid chromatography (UPLC).

The obtained fractions were filtered using a 0.45 μm membrane in a Millex filtration system (Millipore) and subsequently subjected to a second 0.22 μm membrane in a Millex filtration system (Millipore). After that, each sample was diluted 20 times with purified water using the Milli-Q water purification system (Millipore).

The prepared samples were analyzed by UPLC using an Agilent 1290 Series chromatograph (Agilent Technologies) configured with a G4204A quaternary pump, G4226A autoinjector, column oven, and DAD detector with fixed detection wavelengths at 280 and 325 nm. An Acquity UPLC HSS T3 column (Waters) was used.

Chromatographic analyses were performed with a gradient mode. The mobile phase consisted of a mixture of water: acetonitrile and trifluoroacetic acid (0.05% v/v) at a flow rate of 0.5 mL min^{-1} and elution time of 11 min. The column temperature was maintained at 40 °C, and the injection volume was 1.0 μL. Chromatograph control, data acquisition, and processing were performed by EZCrom SI software (Agilent Technologies).

5.2.1 Obtained Results

The results of the lignin cracking reactions were analyzed by UPLC and demonstrated greater representativeness for 5-hydroxymethylfurfural, p-coumaric, and ferulic acids (Fig. 5.5). For the reactions that used Kraft black liquor as a source of lignin, the appearance of other by-products was noted, which is most likely due

Table 5.1 Reaction conditions used in the experiment

Method	Reaction conditions
Standard	2 g of lignin, 20 mL of purified water, $T = 200$ °C, $t = 120$ min
1	2 g of lignin, 20 mL of purified water, $T = 200$ °C, $t = 120$ min, 0.5 g of zeolite catalyst (HZSM-5)
2	2 g of lignin, 20 mL of purified water, $T = 200$ °C, $t = 240$ min
3	2 g of lignin, 20 mL of purified water, $T = 200$ °C, $t = 240$ min, 0.5 g of zeolite catalyst (HZSM-5)
4	2 g of lignin, 20 mL of distilled water, $T = 200$ °C, $t = 240$ min, $P = 90$ psi

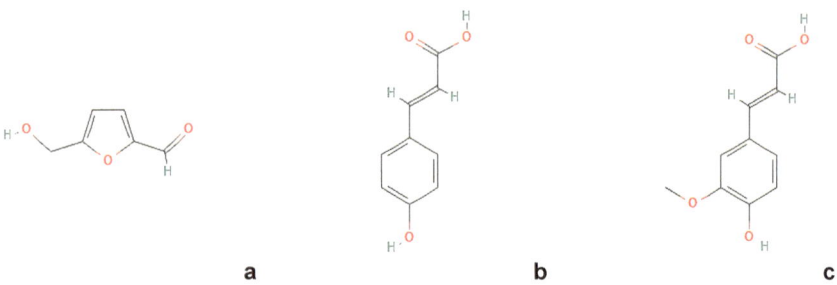

Fig. 5.5 Obtained compounds from lignin cracking: **a** 5-hydroxymethylfurfural; **b** *p*-coumaric acid; and **c** ferulic acid

to the acidic starting medium (pH value of 4.5), which favored the lignin cracking reaction with the emergence of new chromatographic peaks that could not be identified. The other phenolic acids present were in low percentages in relation to the first three compounds, although the chromatograms suggest the predominant presence of oligomers.

It is possible to note the product variability, with different molecular size and chemical groups, which can be a difficult to be overcome in the separation steps (downstream).

5.3   A Critical Evaluation of Processes: Advantages Versus Disadvantages According to Green Chemistry Principles

According to those processes and technologies presented in this chapter—i.e., chemocatalysis, molecular deconstruction (or depolymerization or cracking), thermochemistry, and structural modifications—the first and most important advantaged according to the green chemistry principles [19] is the promotion of the use of renewable feedstock (principle 7), i.e., lignin from lignocellulosic biomass. It is desirable efforts addressed to develop less hazardous chemical synthesis (principle 3) and to design safer chemicals (principle 4).

Unfortunately, regarding to the design for energy efficiency (principle 10) it is the Achilles' heel for the thermochemical processes (pyrolysis and hydrothermal carbonization) due to the high energy demand, which could make it inadequate for some applications.

5.4 Conclusions

The use of chemical conversion approaches to lignin macromolecule can boost its use as renewable carbonaceous raw material for chemicals, fuels, and materials.

Processes based on chemocatalysis, molecular deconstruction (or depolymerization or cracking), thermochemistry, and structural modifications, despite their limitations, can promote the valorization of the lignin macromolecule in order to create new products and new value chains.

References

1. R. Nadányi, A. Ház, A. Lisý, M. Jablonský, I. Šurina, V. Majová, A. Baco, Lignin modifications, applications, and possible market prices. Energies **15**, 6520 (2022). https://doi.org/10.3390/en115186520

2. L. Lin, X. Han, B. Han, S. Yang, Emerging heterogeneous catalysts for biomass conversion: studies of the reaction mechanism. Chem. Soc. Rev. **50**, 11270–11292 (2021). https://doi.org/10.1039/D1CS00039J

3. P. Kaur, G. Singh, S.K. Arya, Tandem catalytic approaches for lignin depolymerization: a review. Biomass Convers. Biorefinery **14**, 6143–6154 (2024). https://doi.org/10.1007/s13399-022-02980-6

4. N. Luo, M. Wang, J. Zhang, H. Liu, F. Wang, Photocatalytic oxidation–hydrogenolysis of lignin β-O-4 models via a dual light wavelength switching strategy. ACS Catal. **6**, 7716–7721 (2016).https://doi.org/10.1021/acscatal.6b02212

5. D. Zhang, X. Zhang, H. Yin, Q. Zheng, L. Ma, S. Li, Y. Zhang, P. Fu, Production of aromatic hydrocarbons from lignin derivatives by catalytic cracking over a SiO_2–Al_2O_3 catalyst. RSC Adv. **13**, 10830–10839 (2023). https://doi.org/10.1039/D3RA00990D

6. P. Sudarsanam, T. Duolikun, P.S. Babu, L. Rokhum, M.R. Johan, Recent developments in selective catalytic conversion of lignin into aromatics and their derivatives. Biomass Convers. Biorefinery **10**, 873–883 (2020). https://doi.org/10.1007/s13399-019-00530-1

7. Z. Luo, C. Liu, A. Radu, A. Radu, D.F. de Waard, Y. Wang, J.T.B. de Bueren, P.D. Kouris, M.D. Boot, J. Xiao, H. Zhang, R. Xiao, J.S. Luterbacher, E.J.M. Hensen, Carbon–carbon bond cleavage for a lignin refinery. Nature Chem. Eng. **1**, 61–72 (2024). https://doi.org/10.1038/s44286-023-00006-0

8. Z. Yang, Z. Xu, M. Feng, J.R. Cort, R. Gieleciak, J. Heyne, B. Yang, Lignin-based jet fuel and its blending effect with conventional jet fuel. Fuel **321**, 124040 (2022). https://doi.org/10.1016/j.fuel.2022.124040

9. M.E. Jazi, G. Narayanan, F. Aghabozorgi, B. Farajidizaji, A. Aghaei, M.A. Kamyabi, C.M. Navarathna, T.E. Mlsna, Structure, chemistry and physicochemistry of lignin for material functionalization. SN Appl. Sci. **1**, 1094 (2019). https://doi.org/10.1007/s42452-019-1126-8

10. B.F.M.L. Gomes, S.V. Júnior, L.V.A. Gurgel, Production of activated carbons from technical lignin as a promising pathway towards carbon emission neutrality for second-generation (2G) ethanol plants. J. Clean. Prod. **450**, 141648 (2024). https://doi.org/10.1016/j.jclepro.2024.141648

11. A.L. Pauline, K. Joseph, Hydrothermal carbonization of organic wastes to carbonaceous solid fuel—a review of mechanisms and process parameters. Fuel **279**, 118472 (2020). https://doi.org/10.1016/j.fuel.2020.118472

12. C.A. Mullen, in *Thermochemical and Catalytic Conversion of Lignin* ed. by N.P. Nghiem, T.H. Kim, C.G. Yoo. Biomass Utilization: Conversion Strategies (Springer, Cham, 2022). https://doi.org/10.1007/978-3-031-05835-6_7

13. J.-M. Ha, K.-R. Hwang, Y.-M. Kim, J. Jae, K.H. Kim, H.W. Lee, J.-Y. Kim, Y.-K. Park, Recent progress in the thermal and catalytic conversion of lignin. Renew. Sustain. Energy Rev. **111**, 422–441 (2019). https://doi.org/10.1016/j.rser.2019.05.034
14. S. Vaz Jr., C.E. de Salvador, M, Innovative structural modification process of Kraft lignin using continuous-flow regime. Sustain. Chem. Pharm. **36**, 101273 (2023). https://doi.org/10.1016/j.scp.2023.101273
15. W. Thielemans, R.P. Wool, Lignin esters for use in unsaturated thermosets: lignin modification and solubility modelling. Biomacromol **6**, 1895–1905 (2005). https://doi.org/10.1021/bm0500345
16. A. Duval, L. Avérous, Mild and controlled lignin methylation with trimethyl phosphate: towards a precise control of lignin functionality. Green Chem. **22**, 1671–1680 (2020). https://doi.org/10.1039/C9GC03890F
17. Y. Pang, Z. Chen, R. Zhao, C. Yi, X. Qiu, Y. Qian, H. Lou, Facile synthesis of easily separated and reusable silver nanoparticles/aminated alkaline lignin composite and its catalytic ability. J. Colloid Interface Sci. **587**, 334–346 (2021). https://doi.org/10.1016/j.jcis.2020.11.113
18. C. Libretti, L.S. Correa, M.A.R. Meier, From waste to resource: advancements in sustainable lignin modification. Green Chem. (2024). https://doi.org/10.1039/D4GC00745J
19. ACS Green Chemistry Institute (2024), https://www.acs.org/greenchemistry/principles/12-principles-of-green-chemistry.html. Accessed April 2024

Chapter 6
Biochemical and Biological Processing and Their Products

Abstract Enzymes and microbe agents (i.e., yeasts, fungi, and bacteria), followed by posterior biosynthesis of bioproducts, are the basis for the biochemical processing of lignin according to industrial biotechnology. It is worth to consider that the main purpose of bioprocesses is to replace raw materials and products of fossil origin by their renewable counterparts derived from biomass and their constituents. Thus, this chapter treats about enzymatic catalysis and microbes application to produce several organic chemicals. Furthermore, the vanillin production by fungi and the green chemistry principles application are also explored.

Keywords Enzymes · Fungi · Bacteria · Yeast · Synthetic biology · OMICs · Microbe factory · Industrial biotechnology

6.1 Description of the Most Common Processes Used in the Industry and Those in Development Based on Biochemical/Biological Transformations

The lignin processing through a bioprocess follows strategies related to the macromolecule degradation by means enzymes and microbes (i.e., yeasts, fungi, and bacteria) and a posterior biosynthesis of bioproducts. Furthermore, the *synthetic biology*—a new biotechnological set for conversion methods—can help the bioprocess approaches. In a general way, these bioprocesses can produce fuels and chemicals.

It is worth to consider that the main purpose of bioprocesses is to replace raw materials and products of fossil origin by their renewable counterparts derived from biomass and its constituents, as observed by Vaz Jr. [1]. Thus, it is expected a "greening" of bioproducts and their bioprocesses.

© The Author(s), under exclusive license to Springer Nature Switzerland AG 2024   67
S. Vaz Jr., *The Lignin Macromolecule*,
SpringerBriefs in Applied Sciences and Technology,
https://doi.org/10.1007/978-3-031-75511-8_6

6.1.1 Biochemical: Enzymatic Catalysis

Enzymatic catalysis plays a role in the lignin bioprocessing considering the macromolecular degradation and the posterior biosynthesis of bioproducts—these enzymes are known as *lignolytic enzymes.*

On this way, the lignin-modifying enzymes (LME) are a robust class of biocatalysts that include lignin peroxidase (LiP; EC[1] 1.11.1.14), manganese peroxidase (MnP; EC 1.11.1.13), laccase (LAC; EC 1.10.3.2), versatile peroxidase (VP; EC 1.11.1.16), and dye decolorizing peroxidase (DyP; EC 1.11.1.19) [2].

According to Sheldon et al. [3], enzymes can catalyze unnatural synthetic reactions in the chemical industry, contributing to *industrial biotechnology.* Furthermore, these biocatalysts can work in the main readily available green solvent, water, a capacity that the chemical industry needs to achieve to be sustainable. Sheldon et al. [3] also observed that more robust enzymes such as proteases, amylases, lipases, laccases, and cellulases can be widely applied in the transformation process industry, in order to overcome technical limitations, such as:

(i) Low yield.
(ii) Low reaction rates.
(iii) Low stability to conditions provided by conventional organic synthesis.

Furthermore, such enzymes can currently be purchased commercially. As an example of the application of enzymatic processes for the valorization of lignin, we can mention the use of the laccase enzyme (EC 1.10.3.2; Fig. 6.1) for the lignin structural modification to bioproducts [4, 5].

Enzymes such as carboxylases and terpene cyclases—terpenes have a molecular structure close to that of lignin precursor alcohols—can catalyze C–C bond formation reactions via the presence of carbocations and intermediate radicals. Hydratases and dehydratases can catalyze addition and elimination reactions [6], in order to remove or to insert chemical groups.

On the other hand, the peroxidase enzyme is directly involved in the metabolic route of lignin biosynthesis in plants, with the involvement of radical polymerization [6]. Furthermore, enzymes recognize chiral centers well, promoting enantiomeric resolution of reaction products, when necessary. It is especially useful for pharmaceuticals.

Regarding to the use of enzymes for the biosynthesis of chemicals, we can see in Fig. 6.2 the biosynthesis of triacylglycerols (TAG) and polyhydroxyalkanoate (PHA), compounds obtained from the lignin as precursor.

As a recent strategy to improve the enzymes performance, we can cite the *immobilization* of ligninolytic enzymes on different nanoengineered support matrices resulting in designing nanobiocatalytic system with intensified catalytic performance and long-term stability for efficient lignocellulosic biomass valorization (e.g., lignin). Enzyme incorporation, e.g., laccase, on magnetic nanostructures additionally facilitates separation, recovery, and reusability of magnetic nanobiocatalysts [9]. For

---

[1] EC = Enzyme Commission numbers, https://enzyme.expasy.org/.

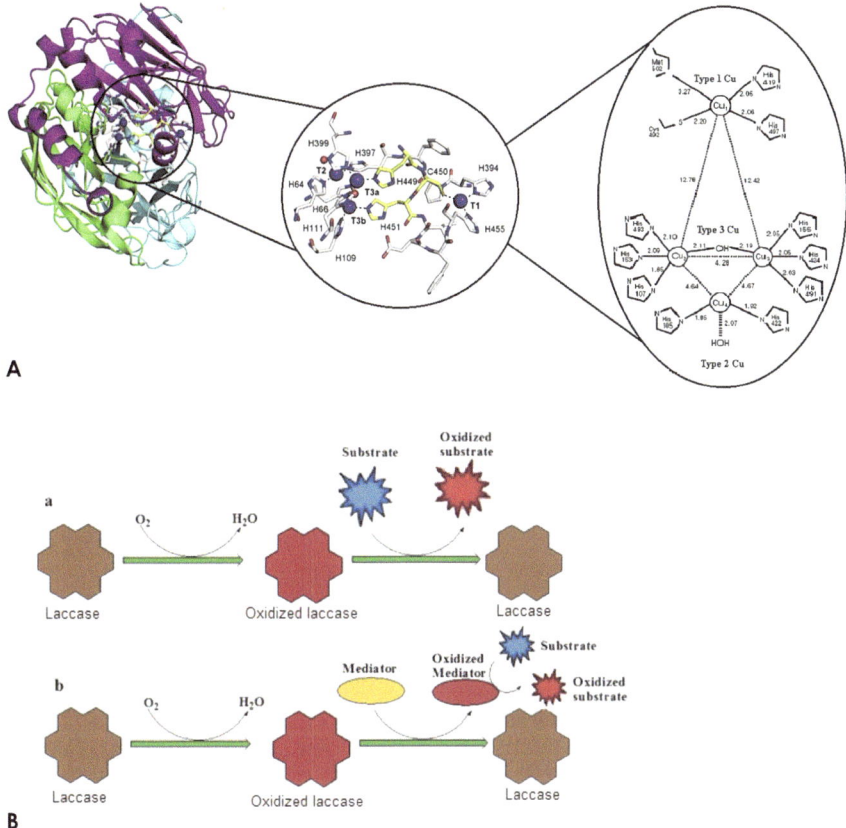

Fig. 6.1 Molecular structure and active site of laccase (**a**); and its catalyzed cycle (**b**). *Source* Adapted from Kumar and Chandra [7]. Reprinted with permission from Elsevier

instance, the use of this same kind of immobilized laccase by magnetic nanoparticles of Fe_3O_4 can improve the enzyme activity and storage stability for the lignin bioconversion [10].

The application of enzymatic catalysis should obey some previous steps depicted in Fig. 6.3.

In a simplified way, firstly the enzyme should be extracted from the microorganism that produce it by means technologies addressed to the cell disruption (intracellular enzymes) or broth concentration (extracellular enzymes). Shortly thereafter, the raw enzyme should be purified by means centrifugation, filtration, ultrafiltration, diafiltration, and chromatography (e.g., ion-exchange, gel filtration, affinity)—purification step is paramount because enzymes are generally found along with other proteins, nucleic acids, polysaccharides, and lipids. After that, the purified enzyme can be immobilized (e.g., in a gel or in a polymer) to improve its performance, stability and

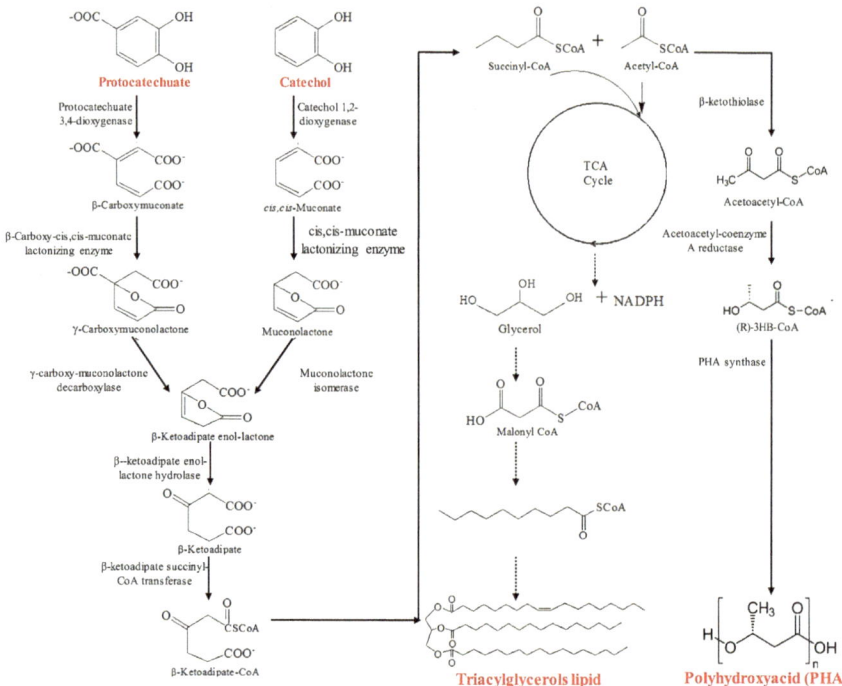

Fig. 6.2 The biosynthesis of triacylglycerols (TAG) and polyhydroxyalkanoate (PHA) from the aromatic ring cleavage of lignin. *Source* Liu et al. [8]. Reprinted with permission from Elsevier

recovery, or directly applied in the bioprocess. And the application of chemical analysis is paramount to guarantee the purity grade and the structure for enzymes, highlighting size exclusion chromatography (to determine the molecular size and polydispersity) [11], X-ray diffractometry (to resolve the molecular three-dimensional structures of proteins at near atomic resolution) [12], and mass spectrometry (to measure the molecular weight of intact proteins and enzyme activities) [13].

Additionally, technological advances as aqueous two-phase system (ATPS) can increase the extraction and purification efficiencies of enzymes due to their versatility, lower cost, process integration capability, and easy scale-up (Nadar et al. 2024).

Finally, Robinson [14] conducted a revision about the bases of *enzymology*, such as classification, structure, kinetics and inhibition, and also provided an overview of industrial applications. From these considerations, we can observe technical–economic advantages from the enzymes uses (e.g., work in aqueous medium at mild conditions and high selectivity).

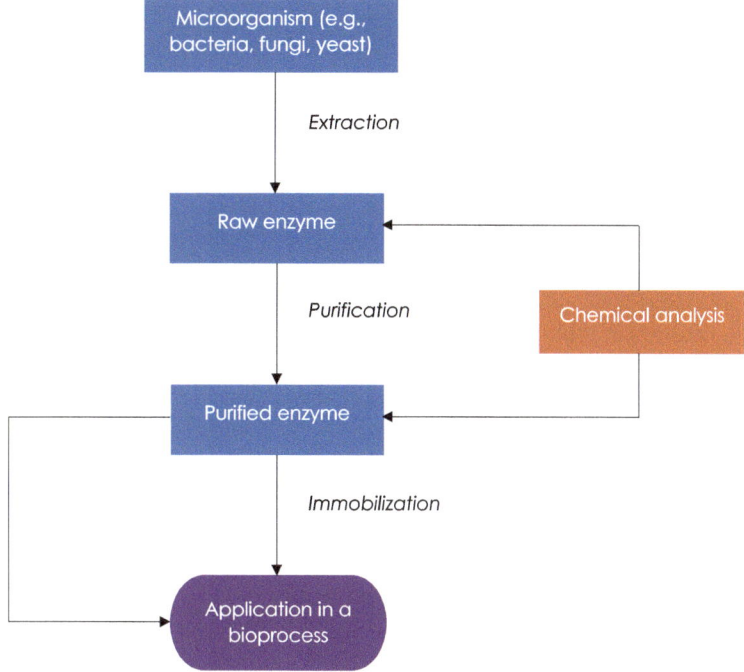

Fig. 6.3 Simplified steps involved in the enzyme application for lignin bioprocessing

6.1.2 Biological: Microbes

Microorganisms play a crucial role in the processing and valorization of lignin, a complex and abundant aromatic macromolecule found in plant cell walls. We can consider here two main strategies:

 (i)   Bacterial-based systems.
 (ii) Fungi-based systems.
(iii) And yeast-based systems.

For bacterial-based systems for lignin valorization we can consider:

- Bacterial cultures have gained attention due to their diverse metabolisms, which can funnel various lignin-based compounds into specific target products or bioproducts.
- Lignin-degrading bacteria are capable of cracking lignin structure and utilizing lignin-associated aromatics.
- These bacteria offer various metabolic pathways that can be harnessed for lignin valorization.

According to the recent advances, researchers have explored bacterial systems for lignin valorization (Fig. 6.4), with some microorganisms can metabolize lignin-derived aromatic compounds, resulting in value-added bioproducts [15]. For instance, engineered microbes with heterologous enzymes can functionalize lignin-derived aromatics into other valuable chemicals. And several bacterial species have been identified for their lignin-degrading capabilities, as *Pseudomonas putida*—known for its lignin-degrading abilities. *Sphingobium sp.*, *Rhodococcus jostii*, and *Rhodococcus opacus* are also commonly studied as promising bacteria for lignin biological processing in order to degrade the macromolecular lignin structure to generate new bioproducts [16].

The extensive scientific literature review and mining depicts that the majority of the microbes for lignin conversion belong to five observed phyla: *Proteobacteria* (114 species/strains), *Basidiomycota* (58 species/strains), *Actinobacteria* (31 species/strains), *Ascomycota* (27 species/strains), and *Firmicutes* (22 species/strains) [16].

The use of bacteria in the lignin fermentation bioprocess can produce [17]:

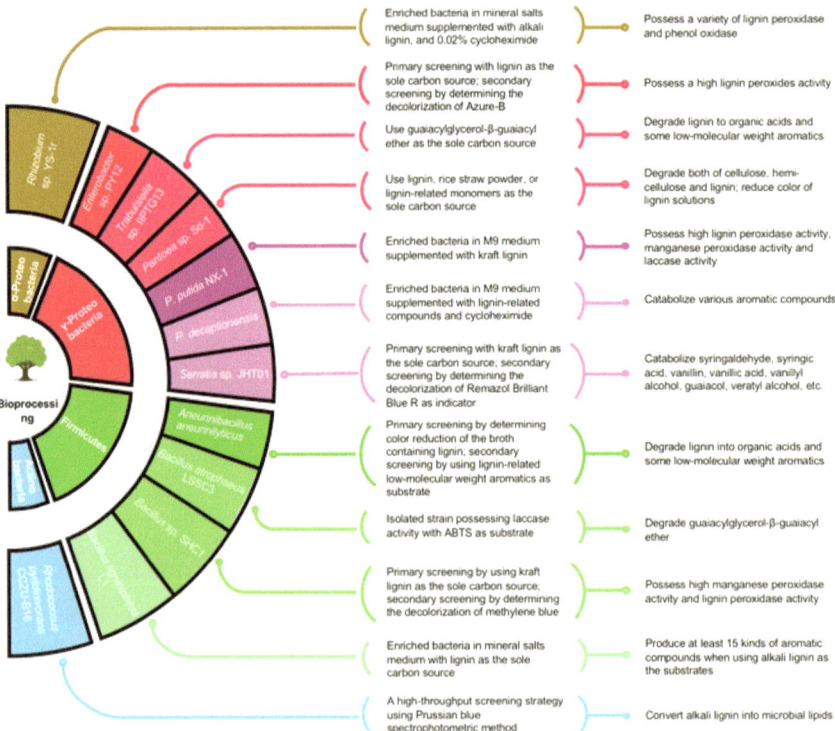

Fig. 6.4 Microbial direct bioprocessing of lignin or lignocellulosic biomass constituents into value-added chemicals through the biological funnel pathway. The lignin-degrading bacteria and their properties are summarized. *Source* Zhang et al. [15]. Reprinted with permission from Science Partner Journal/Nanjing Agricultural University

- Triglycerides (lipids) by *R. opacus.*
- Vanillin by modified *R. jostii* RHA1.
- Polyhydroxyalkanoates by *P. putida* Gpo1 and *P. putida* JCM13063.

Synthetic biology can provide new insights and alternatives for the production of several chemicals by means the design and construction of *microbial factories.* That means, this approach can generate engineered microorganisms using biotechnological tools, as metabolic engineering. It is very useful to modify metabolic pathways in bacteria and yeast in order to produce microbial factories, as proposed by Weiland et al. [18] to reach value-added products from lignin, as "a field of dreams" (Fig. 6.5).

A limitation of microbes—engineered or not—is the high number of by-products that can be generated after the conversion step, which demands additional funds and facilities for the downstream step. And it can be critical for organic chemicals due aspects as optical isomers presence.

OMICs approaches, as *metabolomics* and *genomics*, can be highlighted in order to help the use of synthetic biology and to understand the potential of the bioprocesses. For instance, Moraes et al. [19] applied the OMICs approach to analyze a lignolytic-consortium from sugarcane soil reveling novel genomes and pathways in lignin modification and valorization. Besides, Paul et al. [20] described a comparative interaction profiles of various lignin-degrading enzymes using multi-OMICs, concluding that lignin valorization significantly depends upon metabolic and pathway engineering. The multi-level OMICs components applied to understand the lignin degradation mechanisms embrace:

(i) Genomics (sequencing, assembly, and annotation of the genome, involving *transcriptomics* and *proteomics*).
(ii) Metabolomics (a systematic study of all chemical processes related to metabolites, and it is being used as a potential tool to provide chemical road maps for lignin degradation). Specially for, metabolomics, we can apply analytical tools as nuclear magnetic resonance (e.g., ^{13}C nuclei) and mass spectrometry.

For fungi-based systems for lignin valorization we can take into account:

- Fungi, in particular, are considered ideal microorganisms for the lignin degradation because of their highly effective enzymatic systems.
- Among fungi, white rot fungi (mainly *basidiomycetes*) are prominent lignin degraders.
- They possess lignolytic enzymes such as laccases, lignin peroxidases, and manganese peroxidases that aid in lignin depolymerization [21].

Figure 6.6 show a practical fungi application for lignin valorization, and Fig. 6.7 an example of a basidiomycete fungus—the more promising for this substrate—in nature.

Thus, the use of fungi-based strategy—mainly based on *basidiomycetes* (e.g., white rot fungi or *P. chrysosporium*)—can be more appropriately used to produce enzymes to be extracted and purified, as previously treated in the item 6.1.1.

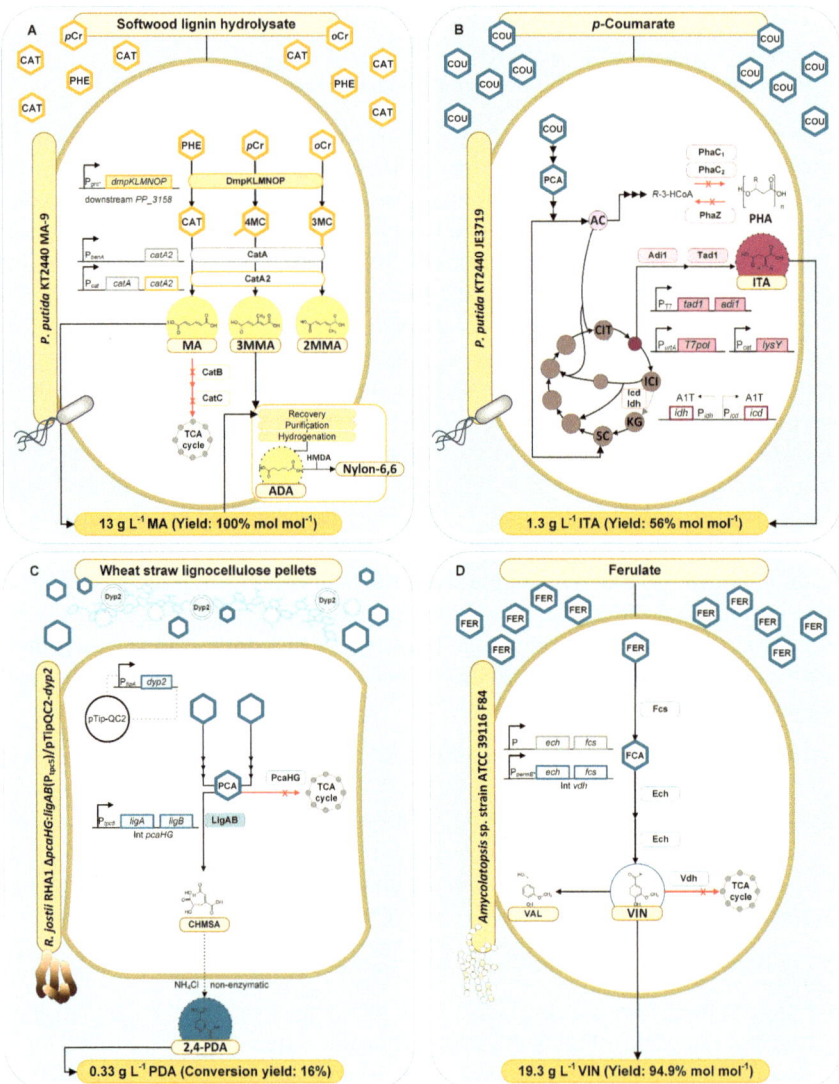

Fig. 6.5 Systems metabolic engineering of different bacterial hosts for the valorization of lignin and lignin-derived aromatic monomers. *Source* Weiland et al. [18]. Reprinted with permission from Elsevier

And to finalize this section, we can consider the use of yeast-based systems. Yeast is a unicellular eukaryotic organism that belongs to the kingdom of Fungi.

For instance, Zhao et al. [22] noticed the production of coumarins—which can act as blood diluting agent and be applied as anti-inflammatory and antioxidant agent—from lignin hydrolysate by means engineered budding yeast, i.e., *Saccharomyces cerevisiae* (Fig. 6.8).

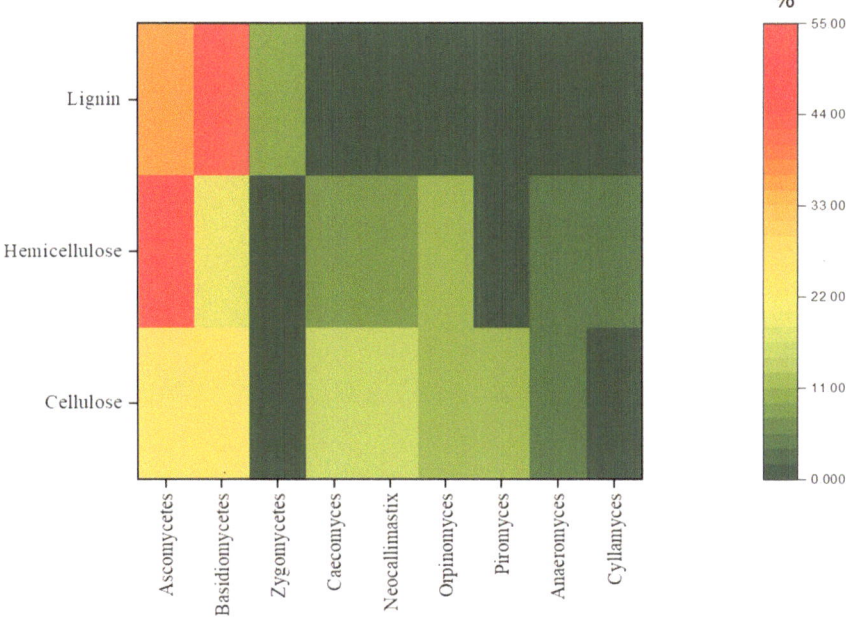

Fig. 6.6 Percentage contribution of fungal families in lignocellulosic biomass degradation. *Source* Dashora et al. [21]. Reprinted with permission from MDPI

Fig. 6.7 Group of white fungi (*Phanerochaete chrysosporium*), a basidiomycete fungus, growing on a rotting log. *Source* Shutterstock

6.2 Discussion of a Case Study

Vanillin (Fig. 6.9)—or bio-vanillin if produced by a bioprocess—is a key aromatic flavor compound used by the industries of foods, beverages, perfumes, and pharmaceuticals. It is typically produced on a large scale through chemical synthesis from

Fig. 6.8 Individual colonies of *Saccharomyces cerevisiae* yeast growing on solid YPD nutrient medium. *Source* Shutterstock

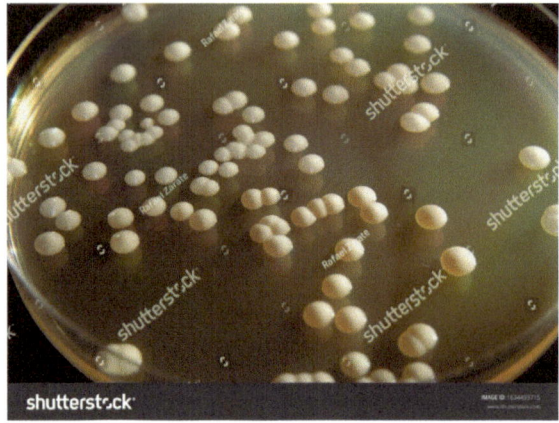

eugenol molecule or extracted from plants (*Vanilla planifolia*). However, alternative biotechnology-based approaches offer promising avenues for vanillin sustainable production.

Regarding to its market forecast [23], we can highlight:

- It is expected to reach US$ 380.7 million by 2033.
- Bio-vanillin manufacturers are offering customized solutions to meet the specific flavor and aroma needs of food and beverage companies, which allows for more tailored and unique product offerings.
- The rise in plant-based and vegan diets has driven the demand for natural flavorings like bio-vanillin. It is used to enhance the taste of plant-based alternatives, including dairy-free ice creams, plant-based cheeses, and non-dairy beverages.
- In the United States, the demand for bio-vanillin is being driven by the increasing popularity of plant-based milk alternatives, such as almond milk and soy milk. Bio-vanillin is often used to enhance the taste of these milk alternatives, which are often perceived as having a less desirable flavor than dairy milk.

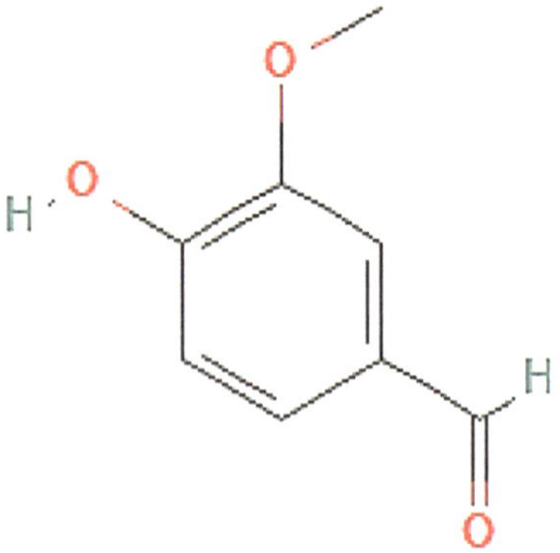

Fig. 6.9 The vanillin molecule or 4-hydroxy-3-methoxybenzaldehyde, a disubstituted phenolic compound

- In India, the demand for bio-vanillin is being driven by the increasing population of vegetarians.

The bioconversion of lignin or its phenolic precursors in vanillin it is an alternative to obtain this chemical by a sustainable route.

Vanillin can be derived from lignin, phenolic stilbenes, isoeugenol, eugenol, and ferulic acid. These chemical precursors serve as starting materials for biotechnological processes. Microorganisms, including fungi, bacteria, and genetically engineered strains [24, 25], can convert these precursors into vanillin. Furthermore, the use of immobilized enzymes—such as attached to a support material—can enhance the efficiency of vanillin synthesis as an alternative to extraction from plants [26].

Despite progress, challenges remain, including precursor accessibility, yield optimization, and pathway control. On the other hand, biotechnological advancements should unlock the potential of sustainable and cost-effective vanillin production. That means, the technological readiness level (TRL) [27] is still below the desired, including bioprocesses and bioproduct under TRL-3 and TRL-4 (validated at laboratory scale) levels.

Unfortunately, and as previously observed by Xu et al. [28], the biotransformation pathways of microbes are inadequate for high vanillin synthesis. Furthermore, it must be taken into account the toxicity of vanillin to microorganism—a characteristic of phenolic compounds. Then, microbes should me engineered to perform an industrial demand with synthetic biology tools.

According these statements, we can see a promise of new bio-based technologies for vanillin production considering an established market, which can valorize the greener of products.

6.3   A Critical Evaluation of Bioprocesses: Advantages Versus Disadvantages According to Green Chemistry Principles

The use a renewable feedstock—according to the green chemistry principle 7 (ACS Green Chemistry Institute 2024) is the main advantage provided by lignin bioconversion processes.

The use of biocatalysts, i.e., enzymes, is close related to the principle 9 that promotes the catalyst use instead of stoichiometric reactants and should be seen as an advantage also.

However, the presence of by-products—an inherent characteristic of biological processes—is a disadvantage because it is against the reduction of derivatives (principle 8).

6.4 Conclusions

The production of chemicals from lignin by means biochemical and biological processing can promote several bioproducts based on the application of enzymes and microbes and advanced tools for industrial biotechnology.

However, this is not industrially established yet instigating the development of new technologies and economically viable bioprocesses based on the "greening" appeal.

References

1. S. Vaz Jr, *Renewable Carbon—Science, Technology and Sustainability* (Elsevier, Amsterdam, 2022). https://doi.org/10.1016/C2021-0-02011-3
2. A.K. Singh, H.M.N. Iqbal, N. Cardullo, V. Muccilli, J. Fernández-Lucas, J.E. Schmidt, T. Jesionowski, M. Bilal, Structural insights, biocatalytic characteristics, and application prospects of lignin-modifying enzymes for sustainable biotechnology. Int. J. Biol. Macromol. **242**, Part 3, 124968 (2023). https://doi.org/10.1016/j.ijbiomac.2023.124968
3. R.A. Sheldon, D. Brady, M.L. Bode, The Hitchhiker's guide to biocatalysis: recent advances in the use of enzymes in organic synthesis. Chem. Sci. **11**, 2587–2605 (2020). https://doi.org/10.1039/C9SC05746C
4. C. Crestini, R. Perazzini, R. Saladino, Oxidative functionalisation of lignin by layer-by-layer immobilised laccases and laccase microcapsules. Appl. Catal. A **372**, 115–123 (2010). https://doi.org/10.1016/j.apcata.2009.10.012
5. V. Hämäläinen, T. Grönroos, A. Suonpää, M.W. Heikkilä, B. Romein, P. Ihalainen, S. Malandra, K.R. Birikh, Enzymatic processes to unlock the lignin value. Front. Bioeng. Biotechnol. **6**, 20 (2018). https://doi.org/10.3389/fbioe.2018.00020
6. T.D.H. Bugg, *Introduction to Enzyme and Coenzyme Chemistry* (2nd ed.). (Blackwell Publishing, Oxford, 2004). https://doi.org/10.1002/9781444305364
7. A. Kumar, R. Chandra, Ligninolytic enzymes and its mechanisms for degradation of lignocellulosic waste in environment. Heliyon **6**, e03170 (2020). https://doi.org/10.1016/j.heliyon.2020.e03170
8. Z.-H. Liu, R.K. Le, M. Kosa, B. Yang, J. Yuan, A.J. Ragauskas, Identifying and creating pathways to improve biological lignin valorization. Renew. Sustain. Energy Rev. **105**, 349–362 (2019). https://doi.org/10.1016/j.rser.2019.02.009
9. M. Bilal, H.M.N. Iqbal, Nanoengineered ligninolytic enzymes for sustainable lignocellulose biorefinery. Curr. Opinion Green Sustain. Chem. **38**, 100697 (2022). https://doi.org/10.1016/j.cogsc.2022.100697
10. X. Chen, B. He, M. Feng, D. Zhao, J. Sun, Immobilized laccase on magnetic nanoparticles for enhanced lignin model compounds degradation. Chin. J. Chem. Eng. **28**, 2152–2159 (2020). https://doi.org/10.1016/j.cjche.2020.02.028
11. P. Hong, S. Koza, E.S.P. Bouvier, A review size-exclusion chromatography for the analysis of protein biotherapeutics and their aggregates. J. Liq. Chromatogr. Relat. Technol. **35**, 2923–2950 (2012). https://doi.org/10.1080/10826076.2012.743724
12. U.B. Gawas, V.K. Mandrekar, M.S. Majik, in *Chapter 5—Structural Analysis of Proteins Using X-ray Diffraction Technique*, ed. by S.N. Meena, M.M. Naik. Advances in Biological Science Research (Academic Press, Cambridge, 2019), pp. 69–84. https://doi.org/10.1016/B978-0-12-817497-5.00005-7
13. T. de Rond, M. Danielewicz, T. Northen, High throughput screening of enzyme activity with mass spectrometry imaging. Curr. Opin. Biotechnol. **31**, 1–9 (2015). https://doi.org/10.1016/j.copbio.2014.07.008

14. P.K. Robinson, Enzymes: principles and biotechnological applications. Essays Biochem. **59**, 1–41 (2015). https://doi.org/10.1042/bse0590001

15. Y. Zhang, C. Cheng, B. Fu, T. Long, N. He, J. Fan, Z. Xue, A. Chen, J. Yuan, Microbial upcycling of depolymerized lignin into value-added chemicals. BioDesign Res. **6**, 0027 (2024). https://doi.org/10.34133/bdr.0027

16. D.P. Brink, K. Ravi, G. Lidén, M.F. Gorwa-Grauslund, Mapping the diversity of microbial lignin catabolism: experiences from the eLignin database. Appl. Microbiol. Biotechnol. **103**, 3979–4002 (2019). https://doi.org/10.1007/s00253-019-09692-4

17. L. Zhao, J. Zhang, D. Zhao, L. Jia, B. Qin, X. Cao, L. Zang, F. Lu, F. Liu, Biological degradation of lignin: a critical review on progress and perspectives. Ind. Crops Prod. **188**, Part B, 115715 (2022). https://doi.org/10.1016/j.indcrop.2022.115715

18. F. Weiland, M. Kohlstedt, C. Wittmann, Guiding stars to the field of dreams: metabolically engineered pathways and microbial platforms for a sustainable lignin-based industry. Metab. Eng. **71**, 13–41 (2022). https://doi.org/10.1016/j.ymben.2021.11.011

19. E.C. Moraes, T.M. Alvarez, G.F. Persinoti, G. Tomazetto, L.B. Brenelli, D.A.A. Paixão, G.C. Ematsu, J.A. Aricetti, C. Caldana, N. Dixon, T.D.H. Bugg, F.M. Squina, Lignolytic-consortium omics analyses reveal novel genomes and pathways involved in lignin modification and valorization. Biotechnol. Biofuels Bioprod. **11**, 75 (2018). https://doi.org/10.1186/s13068-018-1073-4

20. M. Paul, N.K. Pandey, A. Banerjee, G.K. Shroti, P. Tomer, R.K. Gazara, H. Thatoi, T. Bhaskar, S. Hazra, D. Ghosh, An insight into omics analysis and metabolic pathway engineering of lignin-degrading enzymes for enhanced lignin valorization. Biores. Technol. **379**, 129045 (2023). https://doi.org/10.1016/j.biortech.2023.129045

21. K. Dashora, M. Gattupalli, G.D. Tripathi, Z. Javed, S. Singh, M. Tuohy, P.K. Sarangi, D. Diwan, H.B. Singh, V.K. Gupta, Fungal assisted valorisation of polymeric lignin: mechanism, enzymes and perspectives. Catalysts **13**, 149 (2023). https://doi.org/10.3390/catal13010149

22. C.-H. Zhao, R.-K. Zhang, B. Qiao, B.-Z. Li, Y.-J. Yuan, Engineering budding yeast for the production of coumarins from lignin. Biochem. Eng. J. **160**, 107634 (2020). https://doi.org/10.1016/j.bej.2020.107634

23. Future Market Insights, Bio vanillin market (2023), https://www.futuremarketinsights.com/reports/bio-vanillin-market. Accessed April 2024

24. W. Jiang, X. Chen, Y. Feng, J. Sun, Y. Jiang, W. Zhang, F. Xin, M. Jiang, Current status, challenges, and prospects for the biological production of vanillin. Fermentation **9**, 389 (2023). https://doi.org/10.3390/fermentation9040389

25. D. Zhu, L. Xu, S. Sethupathy, S. Haibing, F. Ahmad, R. Zhang, W. Zhang, B. Yang, J. Sun, Green Chem. **23**, 9554–9570 (2021). https://doi.org/10.1039/D1GC02692E

26. T. Furuya, M. Kuroiwa, K. Kino, Biotechnological production of vanillin using immobilized enzymes. J. Biotechnol. **243**, 25–28 (2017). https://doi.org/10.1016/j.jbiotec.2016.12.021

27. D. Humbird, Expanded technology readiness level (TRL) definitions for the bioeconomy (2018), https://www.biofuelsdigest.com/bdigest/2018/10/01/expanded-technology-readiness-level-trl-definitions-for-the-bioeconomy/. Accessed April 2024

28. L. Xu, F. Liaqat, J. Sun, M.I. Khazi, R. Xie, D. Zhu, Advances in the vanillin synthesis and biotransformation: a review. Renew. Sustain. Energy Rev. **189**, Part A, 113905 (2024). https://doi.org/10.1016/j.rser.2023.113905

Chapter 7
Chemical and Physical Processing for Materials and Their Products

Abstract Carbon-based materials obtained from lignins are a subject to be explored in order to create new high-value products for pulp and paper and biofuel industries. New products as battery components, polymeric materials, additive for pellets, and carbon fiber are presented and discussed in this chapter in order to demonstrate the technological opportunities for these lignin-based materials. Furthermore, a case study will provide the potential of carbon fibers, and the evaluation of advantages and disadvantages according green chemistry principles.

Keywords Carbon lignin-based materials · Polymers · Carbon fibers · Additives · Batteries

7.1 Description of the Most Common Processes Used in the Industry and Those in Development Based on Physical and Chemical Technologies for Materials

Technologies addressed to lignin processing for material use can involve those technologies which necessarily don't need of chemical reactions or biochemical transformation pathways and seen in Chaps. 5 and 6 for chemical and biochemical/biological conversion processes, respectively. That means, we expected a reduced investment demand related to CAPEX[1] and OPEX.[2] Additionally, we can consider some applications addressed to materials use by means a physical processing (e.g., molding by increasing temperature, extrusion, pressing) allied to chemical processing (e.g., polymerization).

Puziy et al. [1] reviewed several carbon lignin-based materials, their applications and main manufacturing processes. They observed that alternatively, technical lignin could be used for the production of bulky carbon materials like graphitic carbons

---

[1] CapEx = capital expenditure, How to Calculate CapEx—Formula (corporatefinanceinstitute.com).

[2] OpEx = operational expenditure, Operating Expenses—Definition, Example, Type, Explain (corporatefinanceinstitute.com).

© The Author(s), under exclusive license to Springer Nature Switzerland AG 2024   81
S. Vaz Jr., *The Lignin Macromolecule*,
SpringerBriefs in Applied Sciences and Technology,
https://doi.org/10.1007/978-3-031-75511-8_7

for carbon composites, glassy carbon for electrodes or sensors, and carbon black for nonelectrical applications.

We will consider some highlighted examples of lignin-based materials in this Chapter, in order to introduce their processes and applications.

7.1.1 Materials for Energy Generation

Lignin can be used to produce batteries by means of lignin-derived anode materials (Li-ion batteries anodes and negative electrodes in Na^+ and K^+ batteries) and lignin-derived cathode materials (e.g., Li–S batteries cathodes), as stated by Beaucamp et al. [2]. Additionally, it was noted that for energy storage in general (batteries and supercapacitors), optimization of electrode architecture is extremely important for maximized electrochemical performance with emerging emphasis on interconnected pore networks, surface area, lignin carbon nanofibers design features (diameter, length, and network rigidity), density/areal loading, and the adoption of cost-effective production strategies suitable for scalability such as 3D patterning technologies.

As an industrial example, we can cite the efforts from Stora Enso and Northvolt to develop wood-based batteries using lignin-based hard carbon made from wood. Their aim is to develop industrialized battery featuring anode sourced from European raw materials, lowering carbon footprint and the cost [3].

Figure 7.1 depicts steps involved in the production of lignin-based electrodes. Certainly, the use of lignin for energic materials is one of the most noble uses.

7.1.2 Polymeric Materials

Researchers have been exploring ways to utilize lignin to produce various polymeric materials, including:

- **Thermosets**: These are durable, heat-resistant materials used in applications like adhesives, coatings, and molded parts.
- **Thermoplastics**: These materials can be melted and reshaped multiple times, making them versatile for manufacturing (Fig. 7.2).
- **Foams**: Lignin-based foams can find applications in insulation and packaging.
- **Hydrogels**: These water-absorbent materials have applications in drug delivery and wound healing.
- **Rubbers**: Lignin-derived rubbers could be used in tires and other elastomeric products.

These polymeric materials are well explored by Zhou et al. [5] in order to achieve rigid-and-flexible, degradable, fully bio-based thermosets, by Parit and Jiang [6] for the potential use of lignin as a component in thermoplastic polymers, as co-polymers

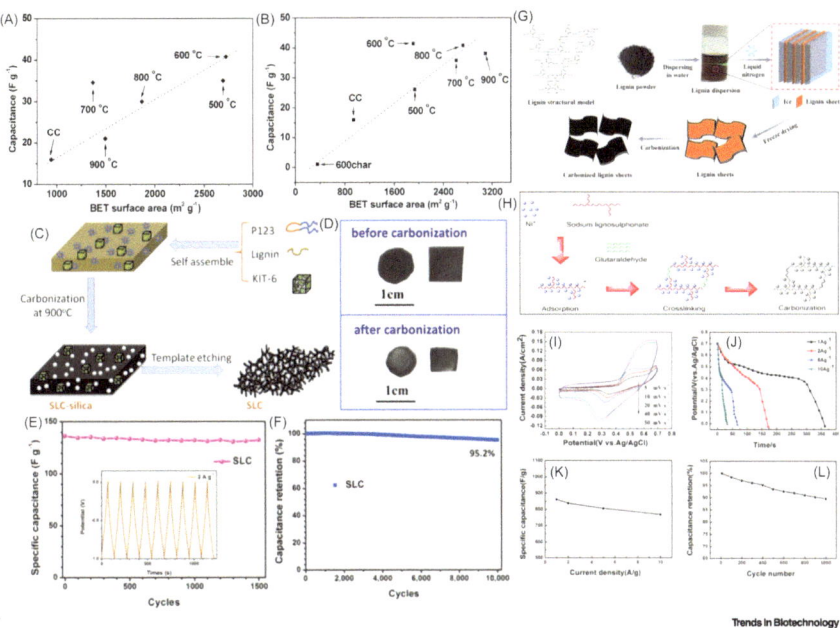

Fig. 7.1 Effect of **a** carbonization and **b** activation temperatures on the capacitance of activated carbon for electrochemical energy systems. *Source* Jia et al. [4]. Reprinted with permission from Elsevier. *SLC* mesoporous-type silica with high surface area and small pores, *KIT-6* mesoporous-type silica with cylindrical pore system; *P123* Pluronics®-type amphiphilic triblock copolymer

and blends; by Adebayo et al. [7] for the development and characterization of lignin-furan rigid foams (Fig. 7.3); by Mondal et al. [8] for the lignin-based hydrogels applications as supercapacitor, biomedical, dye adsorption, and moist induced power generator, and by Sekar et al. [9] for a bio-based hydrothermally treated (HTT) lignin as a potential functional filler for a solution styrene butadiene and butadiene rubber blend.

Some of these new products can reach the technology readiness level (TRL) 6. That means, they are available at demonstration plant [10].

7.1.3   Additive for Improvement of Mechanical Properties of Pellets

Pellets have become an important renewable energy source Barbosa et al. [12]. Aiming to contribute for diversifying the Brazilian energy matrix, the goals of this

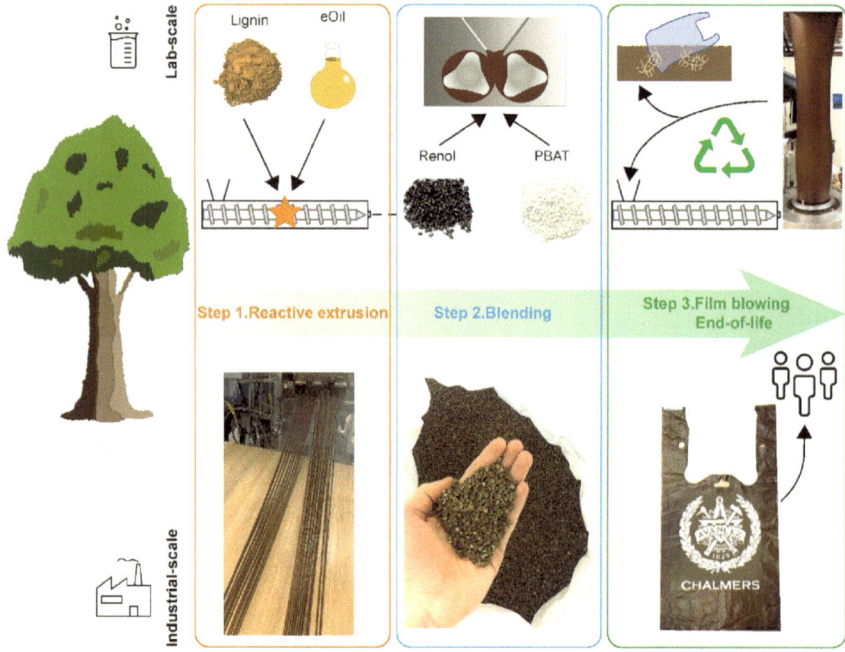

Fig. 7.2 Scheme of the multistep strategy developed for lignin-based thermoplastic: (step 1) reactive extrusion of lignin with methyl 9,10-epoxystearate to produce Renol (a lignin-derived input synthesized by reactive extrusion after manual premixing of lignin with methyl 9,10-epoxystearate (eOil)); (step 2) melt processing of Renol with poly(butylene adipate-co-terephthalate) (PBAT) in an internal mixer; and (step 3) film blowing of Renol/PBAT blends and their recycling. Industrial scale corresponding steps for the production of the biomaterials and a sample of a shopping bag (bottom side). The bag was produced at an industrial facility operating at a production rate of 50 kg h^{-1} and film-blown with a die temperature of 160 °C. *Source* Avella et al. [11]. Reprinted with permissions from Elsevier

work were to evaluate the quality of the pellets of lignocellulosic residues (Eucalyptus and corn) produced with the addition of different percentages of Kraft lignin. For the production of pellets, mixtures of wood with bark of a *Eucalyptus urophylla* and *Eucalyptus grandis*, and corn residue were used as raw material. The proportions of corn residue in the mixture were 0, 20, 25, and 30% (wt/wt). Except for the control (0% lignin), 2 and 5% (wt/wt) Kraft lignin were added to a dry mass of raw material in the four different mixtures (Fig. 7.4). Pellets were produced in a laboratory press pelletizer with horizontal circular array.

The following properties of the pellets were evaluated:

  (i)   Proximate analysis.
 (ii)   High heating value (HHV).
(iii)   Elementary analysis.
 (iv)   Energy density.
  (v)   Bulk density.

Fig. 7.3 Lignin-furan foams. *Source* Adebayo et al. (2023). Reprinted with permission from Springer Nature

  (vi)   Diameter and length.
 (vii)   Hardness.
(viii)   Mechanical durability.
  (ix)   And fine content.

The pellets were classified according to European marketing standards.

The addition of Kraft lignin to eucalyptus and corn residue pellets contributed to improving the physical and mechanical pellet properties, as regards the bulk density, mechanical durability, and fine content, allowing the transportation of a greater amount of mass and energy, besides maintaining the integrity of the biofuels

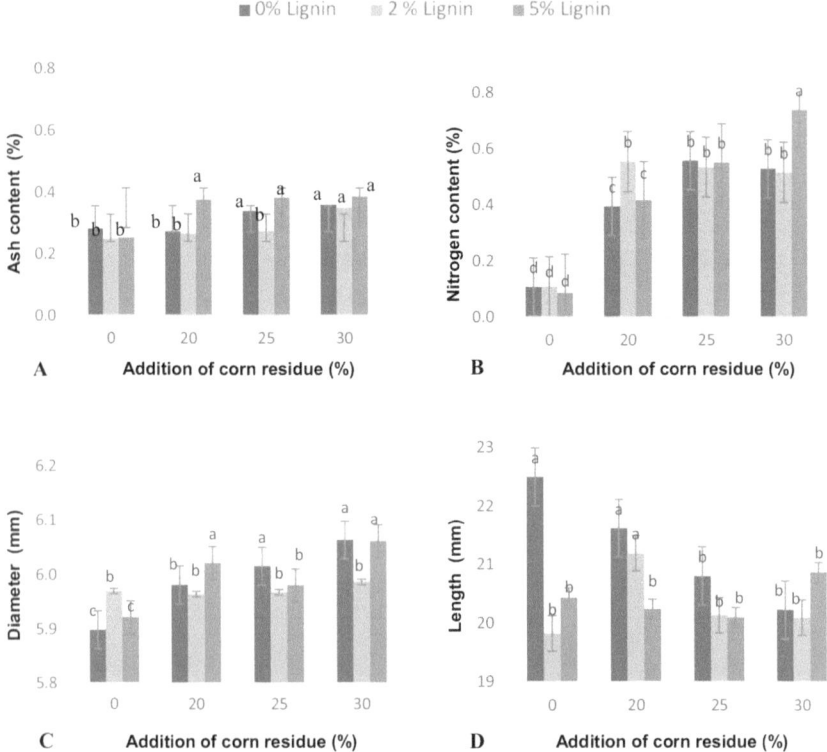

Fig. 7.4 Properties of pellets for the treatments: **a** ash content (%); **b** nitrogen content (%); **c** diameter (mm); **d** mean length of the pellets (mm). Bars followed by the same small letter do not differ among themselves at 5% probability (Scott, $p > 0.05$). *Source* Barbosa et al. [12]. Reprinted with permission from Springer Nature

during handling and use. The mixing of eucalyptus with corn residue is an effective way to optimize properties of biomass solid fuel.

The treatment with higher corn addition, in relation mechanical properties, showed a better performance in accordance with the European standards. The mechanical properties were above to 97.5%, besides that has no impact from the addition of Kraft lignin. The addition of up to 20% of corn residue has the potential to improve physical and mechanical pellet quality, with or without Kraft lignin addition. Thus, similar amounts to that of the treatment with the proportion of 80% eucalyptus and 20% corn residue can be a viable alternative to the production of pellets.

Indeed, the improvement in the mechanical properties can expand the energetic use of pellets and spreading this renewable biofuel.

7.2   Discussion of a Case Study

Carbon fibers (Fig. 7.5) are known for their exceptional strength and stiffness, making them highly desirable for use in renewable composites. Traditionally, carbon fibers are produced from petroleum-derived polymers such as polyacrylonitrile (PAN). However, there is growing interest in finding alternative as sustainable and renewable precursors for carbon fibers.

One such candidate is lignin, as blending lignin with cellulose (e.g., softwood Kraft pulp) can create precursor fibers suitable for carbon fiber production.

The conventional process for making carbon fibers involves two main steps: oxidative stabilization and carbonization. In the stabilization step, precursor fibers are heated to temperatures between 200 and 350 °C to remove volatile components and create a stable structure—lignin-based precursor fibers can be stabilized more rapidly (in less than 2 h) without compromising fiber integrity. After stabilization, the fibers are carbonized at temperatures exceeding 1000 °C to convert them into carbon fibers.

Lignin-derived carbon fibers exhibit several desirable properties:

- **High specific surface area**: beneficial for applications such as energy storage.
- **High porosity**: useful for filtration and catalysis.
- **Good mechanical properties**: although impregnation with ammonium dihydrogen phosphate can increase yield, it may affect mechanical strength.
- **High oxidation resistance and thermal stability**: suitable for high-temperature applications.

This use of lignin as raw material for carbon fibers and nanofibers open new opportunities of application in material engineering. And manufacturing processes, as electrospinning of lignin, can help to obtain these (nano)fibers for several applications [13].

Fig. 7.5 A piece sample of carbon fiber for automotive engineering, obtained from Organosolv lignin from the second-generation ethanol production

7.3   A Critical Evaluation of Processes: Advantages Versus Disadvantages According to Green Chemistry Principles

According to the 12 principles of green chemistry (ACS Green Chemistry Institute 2024), the several uses of lignin for materials can contribute to prevent waste (the principle 1) generated, for instance, by the pulp and paper industry. It can be followed by the promotion of the processing of renewable feedstocks (principle 7). Thus, these approaches can define advantages and environmental gains.

Unfortunately, and specially for polymers, the large use of organic solvents—sometimes with toxic characteristic—cannot contribute to, for instance, the design of beginning chemicals/materials (principle 4) or reducing of derivatives (principle 8). These aspects should be revised in order to produce sustainable materials.

7.4   Conclusions

The use of lignins to obtain carbon-based renewable materials is strategic to develop high-value products for pulp and paper and second-generation ethanol.

Battery components, polymeric materials (e.g., thermosets, thermoplastics, foams, hydrogels, rubbers), additive for pellets, and carbon fiber are excellent examples of new products to be explored, under a holistic view related to green chemistry and considering sustainability by the environmental gains.

References

1. A.M. Puziy, O.I. Poddubnaya, O. Sevastyanova, Carbon materials from technical lignins: recent advances. Top. Curr. Chem. **376**, 33 (2018). https://doi.org/10.1007/s41061-018-0210-7
2. A. Beaucamp, M. Muddasar, I.S. Amiinu, M.M. Leite, M. Culebras, K. Latha, M.C. Gutiérrez, D. Rodriguez-Padron, F. del Monte, T. Kennedy, K.M. Ryan, L. Rafael, M.-M. Titirici, M.N. Collins, Lignin for energy applications—state of the art, life cycle, technoeconomic analysis and future trends. Green Chem. **24**, 8193–8226 (2022). https://doi.org/10.1039/D2GC02724K
3. Industry Intelligence, Inc. Stora Enso, Northvolt partner to develop wood-based batteries using lignin-based hard carbon… *Press release* (2022), https://www.industryintel.com/research-and-development-or-patents/news/stora-enso-northvolt-partner-to-develop-wood-based-batteries-using-lignin-based-hard-carbon-made-with-wood-from-nordic-forests-aim-is-to-develop-industrialized-battery-featuring-anode-sourced-from-european-raw-materials-lowering-carbon-footprint-cost-157481372688. Accessed April 2024
4. R. Jia, C. He, Q. Li, S.-Y. Liu, G. Liao, Renewable plant-derived lignin for electrochemical energy systems. Trends Biotechnol. **40**, 1425–1438 (2022). https://doi.org/10.1016/j.tibtech.2022.07.017
5. S. Zhou, K. Huang, X. Xu, B. Wang, W. Zhang, Y. Su, K. Hu, C. Zhang, J. Zhu, G. Weng, S. Ma, Rigid-and-flexible, degradable, fully biobased thermosets from lignin and soybean oil:

synthesis and properties. ACS Sustain. Chem. Eng. **11**, 3466–3473 (2023). https://doi.org/10.1021/acssuschemeng.2c06990

6. M. Parit, Z. Jiang, Towards lignin derived thermoplastic polymers. Int. J. Biol. Macromol. **165**, Part B, 3180–3197 (2020). https://doi.org/10.1016/j.ijbiomac.2020.09.173

7. A.J. Adebayo, O.O. Oluwasina, J.K. Ogunjobi, L. Lajide, Development and characterization of lignin-furan rigid foams by varying precursors and catalyst concentration. Int. J. Environ. Sci. Technol. **21**, 3087–3102 (2024). https://doi.org/10.1007/s13762-023-05164-5

8. A.K. Mondal, M.T. Uddin, S.M.A. Sujan, Z. Tang, D. Alemu, H.A. Begum, J. Li, J. Huang, Y. Ni, Preparation of lignin-based hydrogels, their properties and applications. Int. J. Biol. Macromol. **245**, 125580 (2023). https://doi.org/10.1016/j.ijbiomac.2023.125580

9. P. Sekar, J.W.M. Noordermeer, R. Anyszka, H. Gojzewski, J. Podschun, A. Blume, Hydrothermally treated lignin as a sustainable biobased filler for rubber compounds. ACS Appl. Polym. Mater. **5**, 2501–2512 (2023). https://doi.org/10.1021/acsapm.2c02170

10. G.A. Buchner, K.J. Stepputat, A.W. Zimmermann, R. Schomäcker, Specifying technology readiness levels for the chemical industry. Ind. Eng. Chem. Res. **58**, 6957–6969 (2019). https://doi.org/10.1021/acs.iecr.8b05693

11. A. Avella, M. Ruda, C. Gioia, V. Sessini, T. Roulin, C. Carrick, J. Verendel, G.L. Re, Lignin valorization in thermoplastic biomaterials: from reactive melt processing to recyclable and biodegradable packaging. Chem. Eng. J. **463**, 142245 (2023). https://doi.org/10.1016/j.cej.2023.142245

12. B.M. Barbosa, S. Vaz, J.L. Colodette, H.F. de Siqueira, C.M.S. da Silva, W.L. Cândido, Effects of Kraft lignin and corn residue on the production of Eucalyptus pellets. Bioenergy Res. **16**, 484–493 (2023). https://doi.org/10.1007/s12155-022-10465-7

13. R. Yadav, O. Zabihi, S. Fakhrhoseini, H.A. Nazarloo, A. Kiziltas, P. Blanchard, M. Naebe, Lignin derived carbon fiber and nanofiber: manufacturing and applications. Compos. B Eng. **255**, 110613 (2023). https://doi.org/10.1016/j.compositesb.2023.110613

Chapter 8
Aspects of Circularity and Sustainability to be Addressed for the Lignin Uses

Abstract Circularity and sustainability are concepts to be strategically incorporated to the lignin processing and end-use in order to confer environmental, social, and economic gains. In this chapter these concepts are explored, allied to bioeconomy, life cycle assessment, and industrial ecology in order to promote sustainable technologies. Additionally, know the biobased content is paramount to guarantee the origin and the sustainability of products and processes assisted by the E-factor. Furthermore, the decarbonization promoted by the lignin-based carbon fibers are discussed as a case study.

Keywords Life cycle assessment · Decarbonization · Industrial ecology · Bioeconomy · UN sustainable development goals · Biobased content · E-factor

8.1 Introduction to Circularity and Sustainability Concepts and Models

The concept of *circularity* or *circular economy* and its application can contribute to a holistic vision of the economic context of the industrial and agroindustrial residues harnessing, as lignin. The European Parliament [1] defines it as "*a model of production and consumption, which involves sharing, leasing, reusing, repairing, refurbishing and recycling existing materials and products as long as possible. In this way, the life cycle of products is extended. In practice, it implies reducing waste to a minimum. When a product reaches the end of its life, its materials are kept within the economy wherever possible. These can be productively used again and again, thereby creating further value*". This concept can be observed in Fig. 8.1 for lignocellulosic raw material.

Considering these statements, the circular economy concept can be explored for the lignin valorization by means these approaches:

- The creation of new value chains, in order to add value to paper and pulp and second-generation ethanol.

Fig. 8.1 Illustration of the circular economy concept. Raw material (e.g., wood) and its processing (e.g., pulping) will generate a product (e.g., cellulose) and its waste (e.g., lignin) which will be reused as raw material (e.g., now lignin)—if their physicochemical properties are adequate

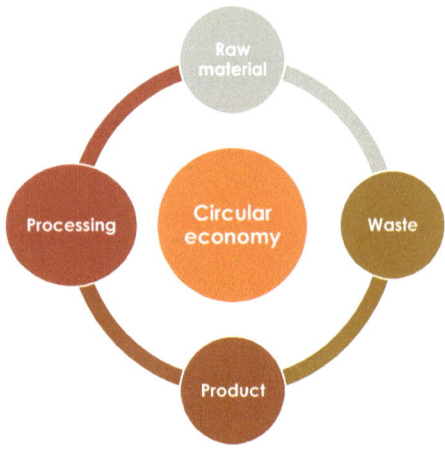

- The reuse and recyclability of materials, as polymers.
- The treatment and mitigation of chemical pollution.

Regarding to *sustainability*, as a theme that call very attention in the twenty-first century by the global society, it is a not so simple concept. The development of sustainable processes and products is a continuous searching in a very technological world that can become a certain good more valuable. Probably the most suitable definition comes from the United Nations [2], as "*meeting the needs of the present without compromising the ability of future generations to meet their own needs*".

Sustainability looks to the future of our resources and life quality by means smart strategies, which involves the 17 Sustainable Development Goals (SDGs) [3]. These SDGs are depicted in Fig. 8.2.

Goals 6 (clean water and sanitation), 7 (affordable and clean energy), 9 (industry, innovation and infrastructure), 12 (responsible consumption and production), and 13 (climate action) are closely related to the lignocellulosic biomass sources and processing in order to achieve sustainable productive chains, as those developed from the lignin uses presented in the previous chapters.

For didactic purposes, the sustainability concept comprises three components or pillars:

- Economic impacts.
- Social impacts.
- Environmental impacts.

These impacts can be *positives* or *negatives* according the appropriate metrics.

The three pillars or components of sustainability are by the time composed by a variety of internal indicators to be considered, as [4]:

- Economic impacts: natural resources use, environmental management, and pollution prevention applied to air, water, land, and waste.
- Social impacts: standard of living, education, community, equal opportunities.

Fig. 8.2 The 17 sustainable development goals (SDGs) [3]. Reprinted with permission from the United Nations

- Economic impacts: profit, cost savings, economic growth, research and development.

The relationship between circularity/circular economy and sustainability can be drawn with the sustainable production chains supporting the firstly.

Furthermore, both concepts can be linked by means a third concept, the *bioeconomy*. This concept is based on the use of biological sources of raw materials to produce economic gains and social beneficiates for the modern society driven by science and technology [5], which is close related to the circular and sustainable uses of a renewable macromolecule from biological sources (i.e., lignin) to promote economic, social, and environmental gains.

8.2 Establishing of Metrics for Sustainability

Several indicators can be listed according the productive chain, turning them suitable to each case of study, i.e., they should attend to the demand. Furthermore, Sikdar [6] suggested consider the intersection—or interdependencies—of the pillars with more internal indicators, what can generate:

- Environmental-economic: energy efficiency, subsides/incentives for use of natural resources.
- Social-environmental: environmental laws, natural resources stewardship.
- Economic-social: business ethics, fair trade, worker's rights.

Then, it is expected a positive impact from all three sustainability components in order to deliver more friendly consumer goods to the society.

When we consider the processing of lignins there are some categories of specific sustainability indicators available for use, mainly for the environmental and social components in order to achieve a reduced environmental footprint and increased societal value [7].

For a reduced environmental footprint:

- Greenhouse gas emissions (GHG): reduction or absence.
- Persistent toxic emissions, as dioxins and furans: reduction or absence.
- Material intensity: reduction.
- Ecological impacts: reduction or absence.
- Land use: reduction.
- Water intensity: reduction or absence.
- Energy intensity: reduction.

Regarding GHG emissions, it is related to carbon dioxide (CO_2) and methane (CH_4) released on the atmosphere. Persistent toxic emissions, for instance, release of dioxins and furans (gaseous effluent). For material intensity, the amount of feedstock used in the processing; moreover, for land use carbon dioxide released by tillage.

For an increased societal value:

- Poverty alleviation: improvement.
- Health and safety improving: improvement.
- Asset recovery: improvement.
- Prosperity and economic resilience: improvement.
- Biodiversity and ecological resilience: improvement.
- Resource conservation: improvement.
- Human dignity: improvement.

As observed by Vaz Jr. [8], to achieve the goals for reduced environmental footprint and increased societal value, it is paramount the investment in research and development and innovation (R&D&I) to create new green technologies, especially for production systems and for transformation processes.

These new technologies should involve the reduction in water, energy, and inputs use allied to cutting-edge scientific assets.

8.3   Carbon Footprint and Life Cycle Assessment of Products and Processes

The quantity of emitted and captured GHG, i.e., carbon dioxide and methane, is a factor to be consider both for products and processes as a balance of mass to estimate the sustainability.

According Fullana et al. [9], *carbon footprint* (CF) is a method used to quantify the amount of GHG emissions associated with a company (*corporate carbon footprint*, CCF) or with the life cycle of an activity or a product/service (*product carbon footprint*, PCF) in order to determine its contribution to climate change. It is a characteristic explored to reach high percentages of marked share.

The carbon footprint (or C-factor) can be calculated by means Eq. 8.1, adapted from Sheldon [10]:

$$C\text{-factor} = kg\ CO_2\ emitted/kg\ product \tag{8.1}$$

Greater the C-factor value, greater the carbon footprint for a certain product.

In order to establish and organize the basis for the *life cycle assessment* (LCA), which includes the CF in its calculations, the International Organization for Standardization (ISO) published the norm 14040: 2006 [11]. This norm is the global standard for LCA.

In order to measure the economic, social, and environmental impacts of technologies for taking advantage of renewable raw materials, robust methodologies are needed to quantify and analyze their gains, especially environmental gains.

LCA is a sustainability tool that assists in decision-making in various industrial sectors, evaluating the positive and negative impacts of technologies, its processes and products. That means the LCA studies the environmental aspects and potential impacts throughout the life of a product—that is, from the "cradle to the grave"—from the acquisition of the raw material, passing through by production, use, and disposal. It is emphasized that resource use, human health and ecology should be the main aspects to be considered during the impact assessment.

Historically, the LCA structure took its first steps in the 1960s in the United States with The Coca-Cola Company generating the first standout scheme when comparing the environmental performance of packaging [12].

Currently, LCA is considered the most efficient method in the study of environmental impacts and has been highly used in sectors of energy generation, polymer production, mining, transport, etc. The main characteristic of an LCA is the possibility of evaluating the environmental impacts of each component of the process, which allows for a more accurate quantification of the impacts of a given route and, consequently, to optimize the proposition of future goals and decision.

When we consider lignocellulosic biomass as raw material source, the land use change needs to be computed because crops generate this class of impact on the environment with the release of greenhouse gas (GHG)—notably carbon dioxide.

8.3.1 Measurement Software

Software are indispensable tools to measure the LCA parameters of impacts due, for instance, the data amount (inputs), their complexity and their results (outputs).

SimaPro[1] is an LCA software that collects, analyzes, and monitors sustainable development data for products and services and is used in applications such as carbon and water footprints, product design, generation of environmental product declarations (see ahead), and definition of key performance indicators. SimaPro is the most used LCA software worldwide, covering users and entities in more than 80 countries, including universities, industries, and companies, in addition to following the recommendations of the norm ISO 14040: 2006 [11].

Moreover, SimaPro stands out for the possibility of carrying out the analysis by different methods, allowing the creation of countless scenarios for a given process and comparing its results.

Eco-Indicator 99 is one of the most used methods in SimaPro based on endpoints. It creates eco indicators that facilitate the modeling of the process, assigning points/ weights to each identified impact, allowing the quantitative comparison of the harm of each impact to the environment.

8.3.2 Environmental Product Declaration

An environmental product declaration (EPD) is an independently verified and registered document that communicates transparent and comparable information about the life cycle environmental impact of products in a credible way. If the EPD is the final report, the foundation of any EPD is an LCA.

An EPD is a so-called type III environmental declaration that is compliant with the ISO 14025 standard [13]. A type III environmental declaration is created and registered in the framework of a program, such as the International EPD System.[2] EPDs registered in the International EPD System are publicly available and free to download.

8.4 Aspects of Industrial Ecology for Production Chains

The concept of *industrial ecology* (IE) emerged in the wake of the environmental movements that intensified in the 1970s and which took shape with the United Nations Environment Program.[3]

We can consider IE as a scientific branch closely related to sustainability because it proposes the development of production systems considering the mimetisation of the dynamic of nature for ecosystems which can promotes positive environmental impacts on those processes and products based on lignocellulosic sources. Moreover, IE is related to the concept of sustainability insofar as it involves closed production

---

[1] https://simapro.com/.

[2] https://www.environdec.com/home.

[3] https://www.undp.org/.

cycles, with reduction or elimination of dependence, for instance, on non-renewable energy sources with such aspects generating positive impacts on the environment, on the local economy and, consequently, on the society.

Recycling and reuse pathways can be explored because they are directly related to circular economy due to the fact that raw materials and products are reused, and such paths being stimulated by the biomass waste and recycling materials use.

Finally, Baldassare et al. [14] studied the IS from two perspectives: circular economy and IE, and found that they are necessary to define an IS cluster.

8.5 How to Measure the Biobased Content

Know the biobased content is paramount the guarantee the origin and the sustainability of products and processes.

Carbon-14, also known as radiocarbon, is a radioactive isotope present in the atmosphere that is absorbed at ground level by living organisms. Living organisms have a known level of carbon-14 while petroleum-derived substances do not have any carbon-14 content. As a result, carbon-14 analysis for biobased testing measures the exact amount of carbon in a material that comes from biomass sources. This measurement is performed according to test methods such as ASTM D6866 using an accelerator mass spectrometer instrument [15].

Biobased product test results are reported as a ratio of biobased organic carbon to total organic carbon and range from zero percent to one hundred percent biobased. The value is the part of the product's organic carbon that is derived from biobased sources.

A product entirely composed of biobased ingredients will yield a result of one hundred percent biobased content while a product that has only petroleum-derived ingredients will result in zero percent biobased content.

This test is performed by laboratories as Beta Analytic[4] in the USA.

8.6 The E-Factor

The E-factor is a parameter introduced by Sheldon [16] in order to promote the pollution prevention from chemical industry (i.e., oil refining, bulk chemicals, fine chemicals, pharmaceuticals), and can be one more tool to prove the environmental gain of renewable raw materials and their products. Furthermore, it is closely related to the carbon footprint and to some green chemistry principles as the use of renewable feedstock (principle 7) and the use of catalysts rather than stoichiometric reactants (principle 9) [17].

The E-factor can be calculated by means Eq. 8.2:

---

[4] https://www.betalabservices.com/biobased/astm-d6866.html.

$$\text{E-factor} = \text{EF} = \sum m(\text{raw materials}) - m(\text{product})/m(\text{product}) \qquad (8.2)$$

The first equation terms address a sum of waste mass generated by the process for a certain product minus the mass of the product. And the second equation term is the mass of this product. In a simple way, greater the E-factor value, less environmental-friendly is the product.

8.7   Discussion of a Case Study

The use of lignin for the *decarbonization* of transportation sector can sound unusual, but is an opportunity to promote technological, economic, and environmental advances.

Lignin, as a by-product abundant in paper industries and emerging biorefineries, presents several exciting approaches for its valorization. One avenue is the production of carbon fiber (CF) from lignin-based carbonaceous raw material—previously introduced in Chap. 7. CF is a lightweight material with exceptional mechanical properties (Fig. 8.3), making it ideal for various applications, including aerospace, automobiles, wind turbines, construction, infrastructure, and sporting goods industries. For instance, lightweight vehicles partially constructed with CF can improve fuel efficiency, contributing to decarbonization—or the reduction or absence of carbon dioxide and methane—in the transportation sector [18].

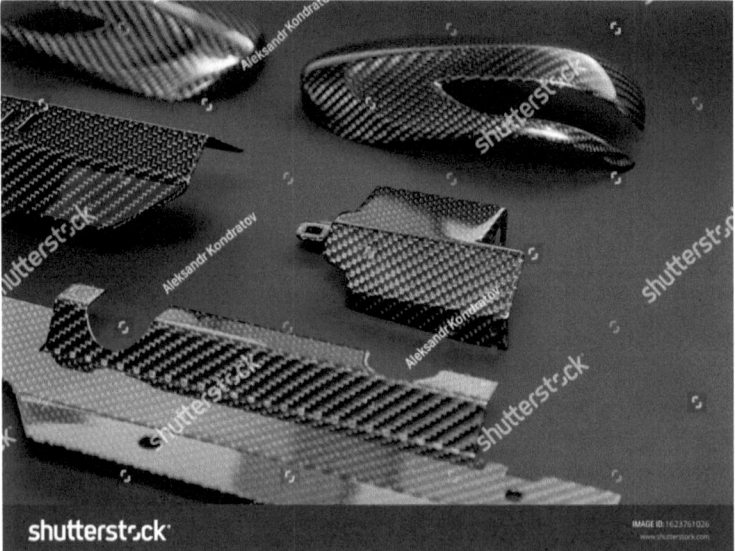

Fig. 8.3   A car exterior elements made from carbon fiber of interwoven black and gray color from heavy-duty yarns for the production of light and durable elements in industry. *Source* Shutterstock

However, the challenge lies in obtaining high-quality lignin-based carbon fibers at a low cost. Previous methods, such as chemical modification or blending with co-precursors, fell short in delivering the desired quality.

A novel approach called *thermo-mechanochemistry* has emerged to overcome this bottleneck. It involves integrating thermal heating and mechanical tension force during the spinning of lignin fibers. This manipulation of ordinary lignin chemistry leads to controlled microstructure evolution.

Using this approach, researchers have successfully produced CF from lignin precursors, achieving adequate tensile strength and modulus values at surprisingly low carbonization temperatures (only 700 °C). This material is ideal for automotive engineering.

8.8 Conclusions

Circularity/circular economy and sustainability are concepts addressed to promote the reuse of materials and to reduce waste associated to the generation of positive environmental, social, and economic impacts. It can be potentialized by means the addition of bioeconomy concept into the recipe.

LCA is considered the most efficient method to measure the environmental impacts by means a holistic view, as using the approach "cradle to the grave" for products and their processes.

Furthermore, the incorporation of aspects of industrial ecology can reinforce the recycling and reuse pathways for raw materials.

Finally, the definition of biobased content, E-factor, and the decarbonization potential can support the sustainability of products.

References

1. European Parliament, Circular economy (2023), https://www.europarl.europa.eu/thinktank/inf ographics/circulareconomy/public/index.html. Accessed April 2024
2. United Nations, Academic impact (2024a), https://www.un.org/en/academic-impact/sustainab ility. Accessed April 2024
3. United Nations, Sustainable development (2024b), https://sdgs.un.org/. Accessed April 2024
4. S. Vaz Jr., in *Introduction to the Sustainability Concept Applied to Renewable Carbon*, ed. by S. Vaz Jr. Renewable Carbon—Science, Technology and Sustainability (Elsevier, Amsterdam, 2022), pp. 167–178. https://doi.org/10.1016/B978-0-323-99735-5.00005-0
5. C. Patermann, A. Aguilar, A bioeconomy for the next decade. EFB Bioecon. J. **1**, 100005 (2021). https://doi.org/10.1016/j.bioeco.2021.100005
6. S.K. Sikdar, Sustainable development and sustainability metrics. AIChE J. **49**, 1928–1932 (2004). https://doi.org/10.1002/aic.690490802
7. United States Environmental Protection Agency, A framework for sustainability indica-tors at EPA (2012), https://www.epa.gov/sites/production/files/2014-10/documents/framew ork-for-sustainability-indicators-at-epa.pdf. Accessed April 2024

8. S. Vaz Jr., in *Basis of Sustainability for Biomass*, ed. by S. Vaz Jr. Treatment of Biomass Residues—A Sustainable Approach (Springer Nature, Cham, 2020), pp. 19–27. https://doi.org/10.1007/978-3-030-58850-2_2

9. P. Fullana, M. Betz, R. Hischier, R. Puig, *Life Cycle Assessment Applications: Results from COST Action 530* (AENOR, Madrid, 2009), https://www.dora.lib4ri.ch/empa/islandora/object/empa%3A7816. Accessed April 2024

10. R.A. Sheldon, The E factor at 30: a passion for pollution prevention. Green Chem. **25**, 1704–1728 (2023). https://doi.org/10.1039/D2GC04747K

11. International Organization for Standardization, ISO 14040:2006. Environmental management—life cycle assessment—principles and frameworks. ISO, Geneva (2006a), https://www.iso.org/standard/37456.html. Accessed April 2024

12. A. Kylili, L. Seduikyte, P.A. Fokaides, in *Life Cycle Analysis of Polyurethane Foam Wastes*, ed. by S. Thomas, A.V. Rane, K. Kanny, V.K. Abitha, M.G. Thomas. Plastics Design Library, Recycling of Polyurethane Foams (William Andrew Publishing, Norwich, 2018), pp. 97–113. https://doi.org/10.1016/B978-0-323-51133-9.00009-7

13. International Organization for Standardization, ISO 14025:2006. Environmental labels and declarations—type III environmental declarations—principles and procedures. ISO, Geneva (2006b), https://www.iso.org/standard/38131.html. Accessed April 2024

14. B. Baldassare, M. Schepers, N. Bocken, E. Cuppen, G. Korevaar, G. Calabretta, Industrial symbiosis: towards a design process for eco-industrial clusters by integrating circular economy and industrial ecology perspectives. J. Clean. Prod. **216**, 446–460 (2019). https://doi.org/10.1016/j.jclepro.2019.01.091

15. ASTM International, Standard test methods for determining the biobased content of solid, liquid, and gaseous samples using radiocarbon analysis (2022), https://www.astm.org/d6866-22.html. Accessed April 2024

16. R.A. Sheldon, The E factor 25 years: the rise of green chemistry and sustainability. Green Chem. **19**, 18–43 (2017). https://doi.org/10.1039/C6GC02157C

17. ACS Green Chemistry Institute, 12 principles of green chemistry (2024), https://www.acs.org/greenchemistry/principles/12-principles-of-green-chemistry.html. Accessed April 2024

18. Y. Luo, M.E.A. Razzaq, W. Qu, A.A.B.A. Mohammed, A. Aui, H. Zobeiri, M.M. Wright, X. Wang, X. Bai, Introducing thermo-mechanochemistry of lignin enabled the production of high-quality low-cost carbon fiber. Green Chem. **26**, 3281–3300 (2024). https://doi.org/10.1039/D3GC04288J

Correction to: Advanced Analytical Techniques for Lignin

Correction to:
Chapter 3 in: S. Vaz Jr., *The Lignin Macromolecule*,
SpringerBriefs in Applied Sciences and Technology,
https://doi.org/10.1007/978-3-031-75511-8_3

The original version of Chapter 3 was inadvertently published before the inclusion of complete titles and references for certain figures and a table, as well as with an incorrect figure representing analytical results, which have now been corrected. The book and the chapter have been updated with the changes.

The updated version of this chapter can be found at
https://doi.org/10.1007/978-3-031-75511-8_3